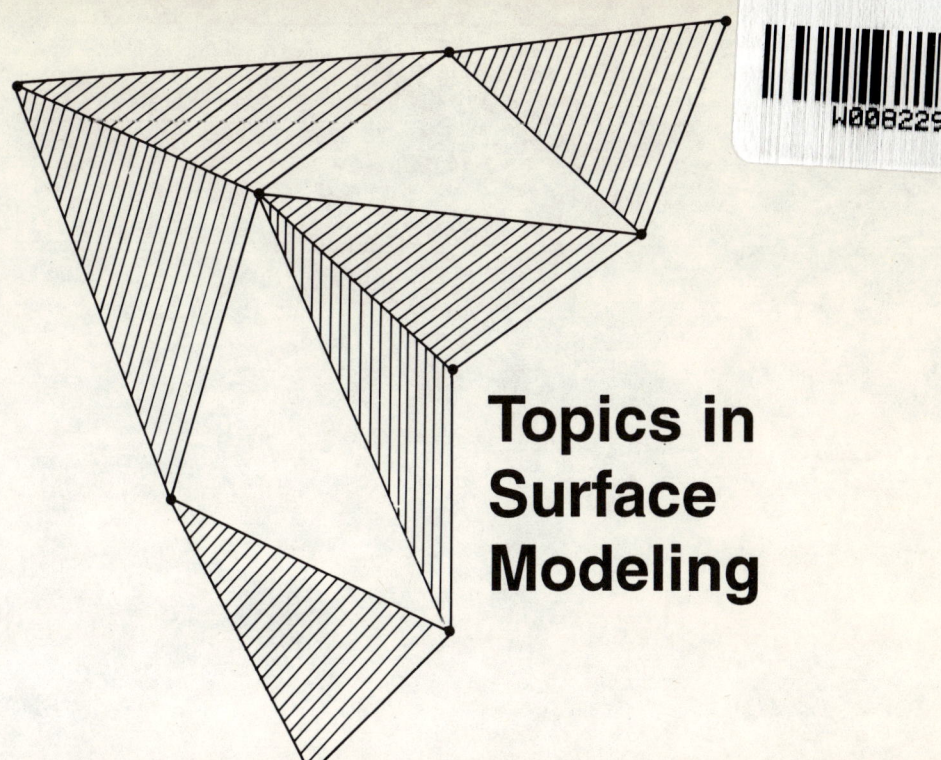

Topics in
Surface
Modeling

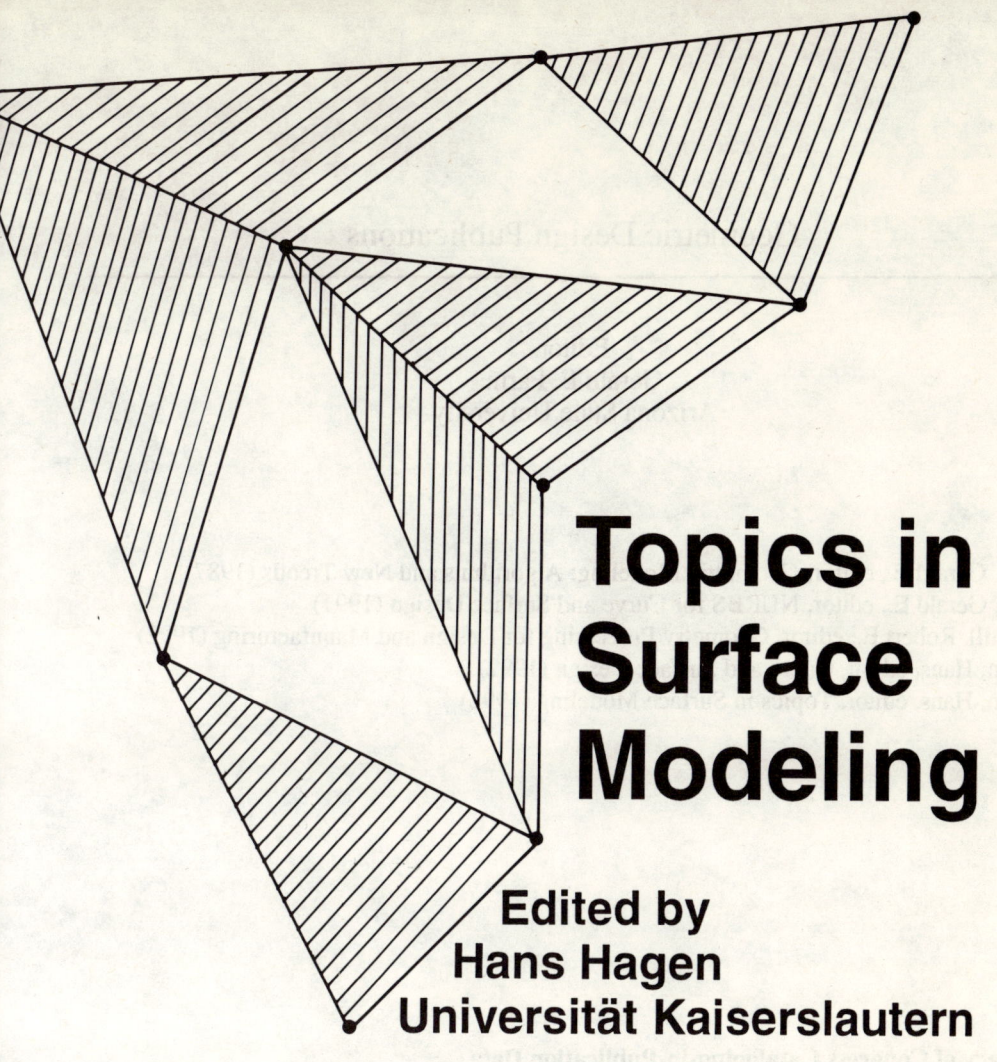

Topics in Surface Modeling

Edited by
Hans Hagen
Universität Kaiserslautern

siam.   Philadelphia

Society for Industrial and Applied Mathematics

Geometric Design Publications

Editor
Gerald E. Farin
Arizona State University

Farin, Gerald E., editor, Geometric Modeling: Algorithms and New Trends (1987)
Farin, Gerald E., editor, NURBS for Curve and Surface Design (1991)
Barnhill, Robert E., editor, Geometry Processing for Design and Manufacturing (1992)
Hagen, Hans, editor, Curve and Surface Design (1992)
Hagen, Hans, editor, Topics in Surface Modeling (1992)

Library of Congress Cataloging-in-Publication Data

Topics in surface modeling / edited by Hans Hagen.
        p.   cm.
        "Sponsored by SIAM Activity Group on Geometric Design"—T. p.
    verso.
        Includes bibliographical references and index.
        ISBN 0-89871-282-3
        1. Surfaces—Mathematical models—Congresses.   2. Curves on
    surfaces—Mathematical models—Congresses.   I. Hagen, H. (Hans),
    1953– .   II. SIAM Activity Group on Geometric Design.
    QA631.T66   1992
    516'.6—dc20   92-12065

Sponsored by SIAM Activity Group on Geometric Design.

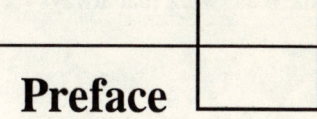

Preface

Curve and surface design methods have grown increasingly sophisticated over the last 30 years. This book contains new ideas and results in three areas of growth: algebraic methods, variational surface design, and special applications. It is intended for practical, industrial applications, as well as research in academic environments.

Part 1. Algebraic Methods. Implicitly defined curves and surfaces have a variety of valuable properties. Here, we present some new ideas and techniques considering practical requirements and real-world examples. Warren's paper contains a free-form blending technique for creating piecewise implicit surfaces. The topics of Bajaj's paper are surface fitting methods and the use of implicit algebraic surface patches. Hoschek and Hartmann use implicit curves and surfaces for interpolation, approximation, and blending of curves, surfaces, and solids, and point out that it is also possible to fill holes in surfaces. The introduced curve and surface design techniques are based on functional spline segments satisfying certain geometric continuity conditions.

Part 2. Variational Surface Design. The generation of "technically" smooth surfaces from a set of three-dimensional data points, appropriate for a milling process, is a key problem in geometric modeling. A new twist estimation method based upon a calculus of variation approach and a stiffness degree concept is presented by Farin and Hagen. This technique creates a smooth surface from a network of curves. In the contribution of Hagen and Santarelli a different point of view is taken. Their variation design method is a combination of a least squares fit and a special smoothing technique, where an initial curve network is not required.

Part 3. Special Applications. Andersson and Dahlberg present interactive techniques for visual design with special emphasis on isophote methods. The paper by Vries-Baayens and Seebregts deals with the problem of converting a trimmed Bézier surface into composite or basic Bézier surfaces.

An important class of surfaces in many applications are functional surfaces such as innerpanels in cars. These surfaces are characterized by highly irregular and multi-featured shapes. Cavendish and Marin describe a new feature-based procedural approach for the design and representation of such functional surfaces. The paper by Jones deals with the contour interpolation problem. Typical applications arise in medical and geological imaging and in computer vision. This paper describes the underlying topological structure of the patchwork to produce topologically correct pseudocontours. This section is concluded by Wassum's report on conditions and constructions for GC^1 and GC^2 continuity between Bézier patches.

This book evolved from presentations given at the SIAM Conference on Geometric Design held in Tempe, Arizona, from November 6–10, 1989. Many chapters began as presentations at that conference. Additionally, certain experts were invited to contribute topics. All submissions were refereed by at least two reviewers, and about half of them were finally accepted for publication in this volume.

I would like to thank Robert E. Barnhill, Gerald Farin, and the SIAM staff for all their effort in organizing the SIAM conference, and especially for their kind help and advice in initiating this book. I would like to express my appreciation for the referees; their support and contribution are always important, but rarely acknowledged. Their names are listed below. Thanks also go to Dieter Lasser, Guido Brunnett, and Ernst Gschwind for numerous and valuable discussions. Last but not least, this book would not have been possible without the countless hours spent by our secretaries Elisabeth Gruys and Karen Lasser on the paperwork that always seems to accompany such an enterprise.

Hans Hagen
Universität Kaiserslautern

Referees

G. Brunnett	T. Foley	D. Lasser
W. Dankwort	G. Farin	G. Nielson
T. DeRose	H. Hagen	D. Roller
R. Franke	J. Hoschek	P. Santarelli

List of Contributors

Roger K. E. Andersson, Department of Mathematics and Statistics, Volvo Data AB , S-40508 Göteborg, Sweden

Chandrajit L. Bajaj, Department of Computer Sciences, Purdue University, West Lafayette, IN 47907

James C. Cavendish, Mathematics Department, General Motors Research Lab., Warren, MI 48090

Björn E. J. Dahlberg, Chalmers University of Technology, Gothenburg, Sweden

Gerald Farin, Department of Computer Science, Arizona State University, Tempe, AZ 85287

Hans Hagen, Universität Kaiserslautern, 6750 Kaiserslautern, Germany

Erich Hartmann, Fachbereich Mathematik, Technische Hochschule Darmstadt, 6100 Darmstadt, Germany

Josef Hoschek, Fachbereich Mathematik, Technische Hochschule Darmstadt, 6100 Darmstadt, Germany

Alan K. Jones, Geometry and Optimization, Boeing Computer Services, Seattle, WA 98124

Samuel P. Marin, Mathematics Department, General Motors Research Lab., Warren, MI 48090

Paolo Santarelli, Steinbeis-Stiftung, TZ RIM, Karlsruhe, Germany

C. H. Seebregts, Faculty of Industrial Design Engineering, Delft University of Technology, 2628 BX Delft, The Netherlands

A.E. Vries-Baayens, Faculty of Industrial Design Engineering, Delft University of Technology, 2628 BX Delft, The Netherlands

Joe Warren, Department of Computer Science, Rice University, Houston, TX 77251.

Peter Wassum, Fachbereich Mathematik, Technische Hochschule Darmstadt, 6100 Darmstadt, Germany

Contents

Contents

Algebraic Methods

Algebraic Methods

Free-Form Blending: A Technique for Creating Piecewise Implicit Surfaces

Joe Warren

1.1. Introduction

Modeling complicated, curved shapes has proven to be a challenging problem in geometric design. One technique that has drawn an increasing amount of interest is the use of implicit surfaces,

$$\{(x, y, z) \mid F(x, y, z) = 0\},$$

to model free-form geometry. Preliminary work by a variety of researchers ([3], [17], [10], [19], [2], [7]) has demonstrated that implicit surfaces have a variety of valuable properties. For example, implicit surfaces naturally bound a volume

$$\{(x, y, z) \mid F(x, y, z) \geq 0\}.$$

Such volumes can be combined in fairly simple ways to form more complicated volumes by combining their defining equations [3], [10]. Surfaces that are fixed distance offsets from polynomial curves and surfaces can be easily specified in implicit form [17], [7].

Creating continuous and tangent plane continuous implicit surfaces is simple. If F is a continuous function (i.e., $F \in C^0$), then $F = 0$ defines a continuous surface. If F is continuous and has continuous first partial derivatives (i.e., $F \in C^1$), then $F = 0$ defines a continuous surface that has a continuous tangent plane almost everywhere. There is no need to maintain and satisfy complicated conditions on a set of control points as in the piecewise parametric case.

However, modeling with implicit surfaces has its problems. Consider the problem of creating a curve that interpolates the points $(1, 1)$ and $(-1, -1)$. Two such curves are $xy - 1 = 0$ and $x^2 - y^2 = 0$, displayed in Fig. 1.1. Neither of these curves is particularly desirable from a modeling viewpoint. The first curve has two separate components with each component containing one of the points. The second curve has a discontinuity in the tangent line at the origin. If implicit surfaces are to be used in serious modeling systems, then many of the shape control techniques that are available for parametric surfaces must be

3

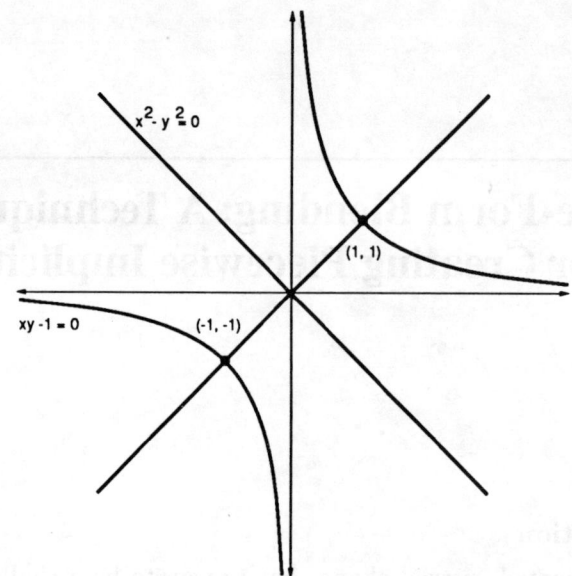

FIG. 1.1. *Interpolating curves may not give pleasing results.*

available for implicit surfaces. In particular, techniques must be available for creating a bounded, connected piece of an implicit surface. The shape of the surface patch should depend only upon local values of input data and should be easily bounded using these input data, in a way that mimics the convex hull property for parametric surfaces.

This chapter describes *free-form blending*, a general technique for creating piecewise implicit surfaces. Given geometric data, input in a variety of forms, free-form blending creates a piecewise implicit surface that approximates these data. Even though the input data may be discontinuous, the method automatically produces implicit surfaces that are C^k continuous for any user specified value of k. The method is local; modifying a piece of the input data affects the shape of the resulting surface only in the vicinity of that data. Free-form blending does not require topological information to accompany the geometric data. For simple variants of the method, versions of the convex hull property and the variation diminishing property hold.

The idea of free-form blending can roughly be summarized as follows:

1. A designer or a computer program provides a collection of geometric data over some region of interest.

2. This region is automatically partitioned into a collection of polyhedral elements.

3. At each vertex of each element, the geometric data at that vertex provide value and derivative data for the first k derivatives of a function.

4. At each vertex, an average value of the C^k data from elements incident to the vertex provides a single piece of C^k data for that vertex.

5. A C^k interpolant defined over the elements produces a C^k interpolating function from these C^k data.

The blended C^k data and C^k interpolant uniquely define an interpolating function $F \in C^k$, and the set $\{(x, y, z) \mid F(x, y, z) = 0\}$, the *zero contour* of F, represents the geometry. Since the interpolating function is C^k, the zero contour that results from interpolating the blended data must be C^k almost everywhere. If the C^k interpolant used is local, then perturbing the data affects the contour only in the vicinity of those data. In most cases, this partition can be a uniform subdivision into regularly shaped elements.

This rest of the chapter discusses in greater detail the key phases of free-form blending. Section 1.2 discusses several interpolants suitable for use with free-form blending and the general properties that such interpolants must have. Section 1.3 describes three natural ways to input C^k data for use with these interpolants. Section 1.4 explores the problem of blending together C^k data in a natural manner and establishes several simple theorems relating the geometry of the input data to the geometry of the implicit surfaces that free-form blending produces. The chapter concludes with several examples of free-form blending applied to various geometric problems. These examples include a simple geometric design application and an application from medical data imaging in which free-form blending has proven particularly powerful.

1.2. Interpolants

In essence, free-form blending is simply an application of some techniques from multivariate interpolation to modeling. An interpolant in its most general form is a method that takes as input a set of data at a collection of points and creates an interpolating function that reproduces the prescribed data at these points. These data may consist not only of values, but also of higher-order derivatives. The interpolants considered in this chapter satisfy the following properties.

Vertex property. The interpolant requires as input a partition of space into polyhedral elements and requires data precisely at the vertices of the elements.

Locality property. The value of the interpolating function along the face of an element, or on the element itself, depends only upon the data values supplied at the vertices of the face.

Subdivision property. For an element with C^k data specified at its vertices, an interpolant produces an interpolating function F. If the element is subdivided into a collection of smaller, similar elements and F provides C^k data values to the newly created vertices, then each interpolating function constructed over a smaller element is identical to F on that element (see Fig. 1.2).

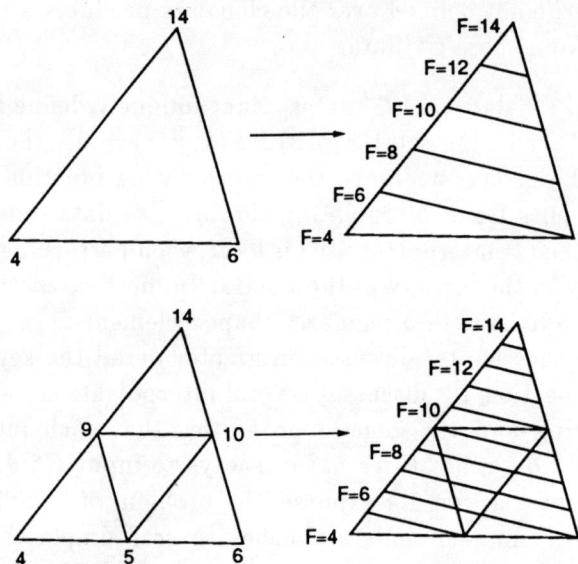

FIG. 1.2. *The subdivision property holds for linear interpolation over a triangular element.*

Locality and subdivision together permit precise shape control for implicit surfaces. Local interpolants ensure that only local data affect the resulting interpolating functions and thus their zero contours. Subdivision for interpolants is the analogue of knot insertion for B-splines. The subdivision property allows a designer to subdivide an element into smaller elements without changing the zero contour, then adjust values in the region of the small elements to make small, local changes to the shape of the contour. The combined properties also permit a data fitting algorithm to adaptively refine an element if the approximation error over the element is too large.

The next few subsections describe various interpolants that exhibit these properties. These interpolants are all piecewise polynomial. Free-form blending can work with other, nonpolynomial interpolants as well. However, piecewise polynomial interpolants are the ones with which the author has had the most experience and the greatest success in implementing free-form blending.

1.2.1. C^0 **Interpolants.** The interpolants considered here can be separated into two distinct classes: simplicial interpolants and cubical interpolants. A *simplex* S in n-dimensions is the convex hull of $n + 1$ points that are not contained in any $n - 1$-dimensional hyperplane. These points are the vertices of the simplex. For example, a two-dimensional simplex is simply a triangle. A three-dimensional simplex is a tetrahedron. If S has vertices $\mathbf{p_0}, \cdots, \mathbf{p_n}$, then the *barycentric* coordinates of a point $\mathbf{q}$ with respect to S are $(\alpha_0, \cdots, \alpha_n)$ where

$$\sum_{0 \leq i \leq n} \alpha_i \mathbf{p_i} = \mathbf{q}, \quad \sum_{0 \leq i \leq n} \alpha_i = 1, \quad \text{and} \quad \bigwedge_{0 \leq i \leq n} \alpha_i \geq 0.$$

Linear interpolation is an example of a simplicial C^0 interpolant. Given the simplex S whose vertices $\mathbf{p_i}$ have values v_i, the unique linear function L that interpolates these values is defined by

$$L(\alpha_0, \cdots, \alpha_n) = \sum_{0 \leq i \leq n} \alpha_i v_i.$$

This interpolant satisfies all three of the properties of §1.2. For example, in two dimensions the subdivision property states that a linear interpolating function over a triangle can also be expressed as four linear interpolating functions over four subtriangles. Figure 1.2 gives an example.

We may also define a C^0 interpolant over a square, a cube or their higher-dimensional analogs. Let $p_{i_1 \cdots i_n}$ be a set of 2^n points in n-dimensions where

$$p_{i_1 \cdots i_n} = (i_1, \cdots, i_n), \quad \text{and} \quad \bigwedge_{1 \leq j \leq n} (i_j = 0 \lor i_j = 1).$$

The convex hull of the points p is called a *hypercube*. *Multilinear interpolation* is an example of a cubical C^0 interpolant. Given a value $v_{i_1 \cdots i_n}$ at the point $p_{i_1 \cdots i_n}$, the unique multilinear function that interpolates the values v is

$$ML(x_1, \cdots, x_n) = \sum_{0 \leq i_1 \leq 1} B^1_{i_1}(x_1) \cdots \sum_{0 \leq i_n \leq 1} B^1_{i_n}(x_n) v_{i_1 \cdots i_n},$$

where $B^1_i(x)$ denotes the first degree Bernstein polynomial $x^i(1 - x)^{1-i}$. Farin [5] gives a nice description of this interpolant for $n = 2$. Again, it is simple to show that this interpolant satisfies all the properties of §1.2.

1.2.2. C^1 **Interpolants.** There are several interpolants that create C^1 interpolating functions from C^1 data. Worsey and Farin [24] describe a piece-wise cubic interpolant that generalizes to n dimensions the two-dimensional simplicial Clough-Tocher interpolant [4]. This interpolant satisfies the first two properties of §1.2, but, unfortunately, it does not possess the subdivision property. The cross-boundary derivatives of the interpolating function along a face are necessarily linear functions, rather than general quadratics, by the conditions that define the interpolant. However, new faces introduced within the simplex during subdivision do not necessarily have linear cross-boundary derivatives. Since the interpolating function chosen depends on the subdivision of an element, this interpolant can only be used if the simplicial mesh is not refined adaptively.

Powell and Sabin describe two C^1 piecewise quadratic simplicial inter-polants for $n = 2$ [15], one consisting of six quadratic functions and the other

consisting of 12 quadratic functions. The 12-way Powell-Sabin split satisfies all
of the properties of §1.2 including the subdivision property. Worsey and Piper
[25] describe a generalization of the six-way Powell-Sabin interpolant to three
dimensions. It is unclear whether this interpolant possesses the subdivision
property.

All of these interpolants are defined for simplicial meshes. In the cubical
case, the interpolants appear to be simpler due to the regular topology
of cubical meshes. To create interpolants for cubical topology, there are
essentially two strategies. The first strategy involves subdivision of the
hypercube into simplices and creation of a piecewise polynomial function over
these simplices. In the two-dimensional case, Sibson [20] describes a piecewise
quadratic and a piecewise cubic C^1 interpolant for the split square. The author
has developed a C^1 piecewise quadratic interpolant for the split cube. This
interpolant satisfies all of the properties of §1.2 as well as quadratic precision.
The examples of free-form blending in Figs. 1.8, 1.9, 1.11, 1.14, and 1.15 were
all made using this interpolant. Generalizing split cube interpolants of this
type to higher dimensions will probably require the application of techniques
from box splines [8], [9].

The second strategy for creating cubical interpolants is to use a tensor
product approach. One such interpolant is a C^1 piecewise multiquadratic
interpolant. This interpolant is a straightforward generalization of the
univariate quadratic interpolant. Although this interpolant can be expressed
as a tensor product B-spline in a manner similar to that of [13], [6], the
following exposition uses the piecewise tensor product Bernstein form. The
shape properties evident from the Bernstein form are useful in proving a few
simple facts concerning the effect of free-form blending. The rest of this section
reviews the multiquadratic interpolant in detail.

In the univariate case, a quadratic function $q(x)$ over the interval $(k, k+1)$
can be represented as a linear combination of the Bernstein basis functions of
order two,

$$q(x) = \sum_{0 \le i \le 2} b_i^k B_i^2(x - k),$$

where $B_i^2(x) = \frac{2}{i!(2-i)!} x^i (1 - x)^{2-i}$. Associated with the coefficient b_i^k is the
domain point $x = k + (i/2)$. Geometrically, b_i^k is an approximation to the value
of q at $x = k + (i/2)$.

Consider two quadratic functions on the intervals $(0, 1)$ and $(1, 2)$, respectively. If the first function has coefficients b_0^0, b_1^0, b_2^0 and the second function
has coefficients b_0^1, b_1^1, b_2^1, then they unite to form a piecewise quadratic function that is continuous if and only if b_2^0 and b_0^1 agree at the common domain
point $x = 1$. The piecewise quadratic function has a continuous first derivative
at $x = 1$ if $b_2^0 = b_0^1$ and the coefficients of the two adjacent domain points
satisfy the linear relation $b_2^0 - b_1^0 = b_1^1 - b_0^1$. Note that the value of $b_2^0 = b_0^1$ can
be derived from b_1^0 and b_1^1 by linear interpolation. Figure 1.3 illustrates this
situation.

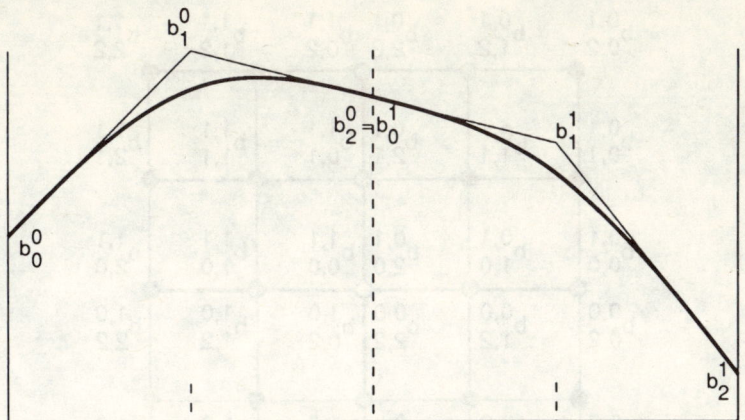

FIG. 1.3. *A univariate quadratic interpolant comes from linear interpolation of control points.*

This piecewise quadratic function can be used to interpolate C^1 data given at the end points of the interval $(0, 2)$. Choose b_0^0 and b_2^1 to match the function values at the end points of the interval, and choose b_1^0 and b_1^1 so that $2(b_1^0 - b_0^0)$ is the derivative at $x = 0$ and $2(b_2^1 - b_1^1)$ is the derivative at $x = 2$. Linear interpolation supplies the remaining value.

The multivariate case proceeds in a similar manner. Shift the unit hypercube so that the origin translates to $K = (k_1, \cdots, k_n)$ and let $q(x_1, \cdots, x_n)$ be a multiquadratic function over that hypercube. The function q can be represented as

$$q(x_1, \cdots, x_n) = \sum_{0 \le i_1 \le 2} B_{i_1}^2(x_1 - k_1) \cdots \sum_{0 \le i_n \le 2} B_{i_n}^2(x_n - k_n) b_{i_1 \cdots i_n}^K.$$

The coefficient $b_{i_1 \cdots i_n}^K$ is associated with the domain point $K + (i_1/2, \cdots, i_n/2)$. Given two multiquadratic functions defined over adjacent hypercubes, the two functions form a piecewise continuous function if and only if the coefficients associated with domain points shared by the hypercubes agree. The two multiquadratic functions form a C^1 piecewise function if and only if all collinear triples of adjacent domain points that span the boundary between the hypercubes have associated coefficients that satisfy a linear function.

Multiquadratics can be used to create a C^1 interpolant. Let C^1 data be specified at the 2^n vertices of the hypercube with opposite vertices $(0, \cdots, 0)$ and $(2, \cdots, 2)$. Partition this hypercube into 2^n subhypercubes and construct over each subhypercube a multiquadratic function in Bernstein basis form.

The C^1 data at a vertex v of the original hypercube define a linear function L_v. Consider the unique subhypercube with edge lengths $1/2$ that contains v. For a domain point p on this subhypercube, let the coefficient associated with p have the value $L_v(p)$. At v, the resulting multiquadratic function interpolates the original C^1 data. The remaining ordinates must be chosen to guarantee

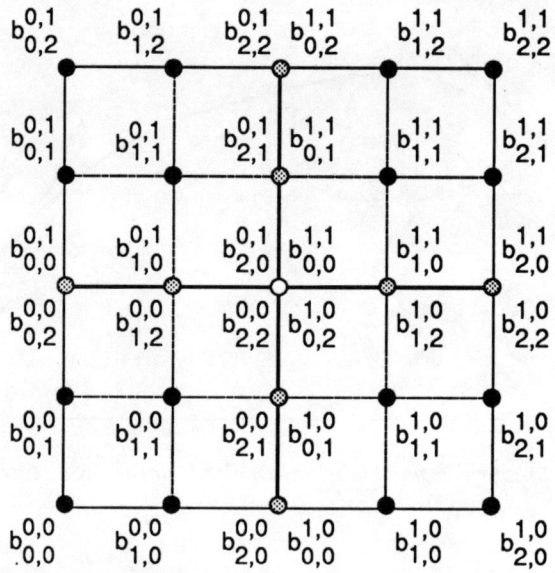

FIG. 1.4. *A biquadratic interpolant on the square.*

C^1 continuity within the hypercube and across its facets. Fortunately, the appropriate values can be calculated by multilinear interpolation. For example, the coefficients of the domain points at the midpoint of an edge of the hypercube are the values that arise from linear interpolation along that edge, as in the univariate case. These values, in turn, are interpolated to provide coefficients of domain points deeper inside the hypercube. Figure 1.4 illustrates this construction in the plane, with lighter circles corresponding to values that are computed by linearly interpolating values marked by darker circles.

The multiquadratic interpolant satisfies all three of the properties enumerated at the beginning of the section. The subdivision property follows directly from the fact that polynomials in Bernstein basis representation can be subdivided using the de Casteljau algorithm.

1.2.3. Higher-Order Interpolants.

Given the existence of C^k interpolants, free-form blending can be used to produce C^k surfaces that approximate some given input data. In the case of simplicial meshes, the author has not had the opportunity to explore some of the C^2 interpolants that are available [1], [18], [23]. One apparent difficulty in the simplicial case is the lack of a comprehensive, general theory of C^k interpolants over simplicial meshes. However, in the case of cubical meshes, the situation appears to be much better.

One possible approach is to subdivide the cubical mesh into a regular simplicial mesh and apply the techniques from box splines [8], [9]. Another approach is to generalize the tensor product approach of the previous section. For example, given C^2 data at the end points of an interval, it is possible to create a C^2 interpolant over that interval using a piecewise polynomial function consisting of three cubic pieces. This interpolant can be generalized via the tensor product approach to create a C^2 piecewise multicubic interpolant over a cubical mesh. More generally, given C^k data at the end points of a line segment, it is possible to create a piecewise polynomial function consisting of $k+1$ polynomial functions of degree $k+1$. Again, generalizing this interpolant to cubical meshes using the tensor product approach is straightforward. Similar techniques have been applied to create multivariate B-splines.

1.3.   Methods for Specifying C^k Data

The strategy in free-form blending is to construct a collection of C^k data at the vertices of a cubical or simplicial mesh. A function interpolating these data is then created using a C^k interpolant. How should the C^k data provided to the interpolant be specified? The most straightforward approach is to have the designer specify the C^k data at the vertices of the mesh directly. However, this approach is impractical in all but a few small cases. The number of data points for even a small mesh would overwhelm most designers. Using these data points to control the shape of the zero contour of the resulting function is also a difficult task. Finally, the interpretation of higher-order data such as second or third derivatives is not very intuitive for most designers.

Nevertheless, such information must be supplied for the resulting interpolation scheme to produce reasonable surfaces. For example, consider the following simple problem. In the univariate case, the designer specifies a sequence of values at the vertices of a sequence of line segments. These values grow linearly. To use a C^1 interpolant to form a C^1 curve that interpolates these values, one must also provide a set of derivatives at each vertex. If one makes the expedient choice of setting these derivatives to be zero, the resulting curve has a zigzag form. For the interpolant to generate a straight line through the points, a derivative equivalent to the slope of the line through the points must be specified at each vertex.

The strategy in free-form blending is to have the designer specify a collection of simple geometric data such as a set of points, a set of lines or planes, or maybe even a collection of higher-order surfaces. These geometric entities are then queried to produce the C^k data necessary for a C^k interpolant. The next three subsections discuss three particularly appealing methods for specifying these geometric data.

1.3.1.   Approximation of Piecewise Linear Functions.   The simplest method is to express the geometric data as a piecewise linear function over the underlying mesh. For example, in the univariate case, the desired function can be specified as a collection of line segments defined over a continuous sequence of intervals. Figure 1.5 illustrates one possible set of input data. Note that the piecewise linear function need not be continuous. From this example, it is clear that the C^1 data collected at the common end point of two adjacent intervals

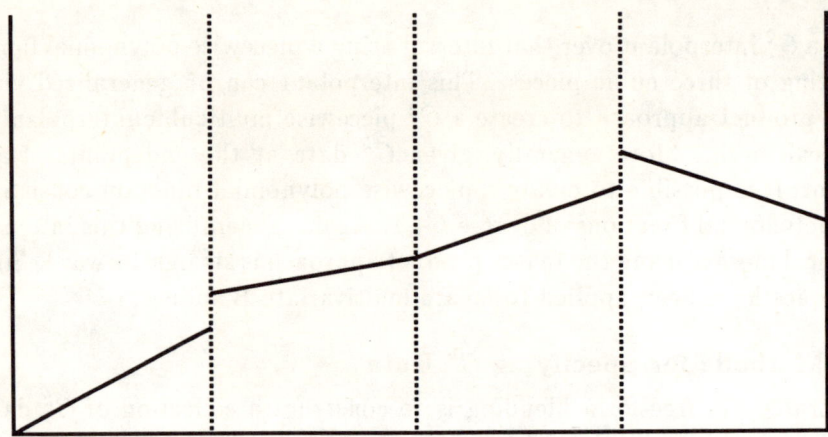

FIG. 1.5. *A piecewise linear function.*

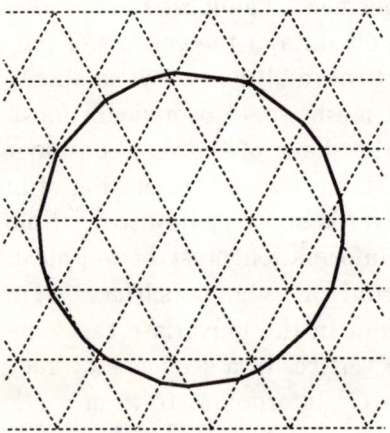

FIG. 1.6. *A circle approximated by line segments on a grid.*

do not necessarily agree. The next section discusses how to reconcile these data. The same technique may be applied in the higher-dimensional case.

To create an implicit curve, one specifies a collection of lines embedded in a mesh that approximate the shape of the curve, as in Fig. 1.6. In the implicit surface case, we specify a set of planes embedded in some three-dimensional mesh. The equation of a given line or plane is only unique up to multiplication by a constant. To normalize the equations of the lines and planes with respect to each other, the simplest approach is to scale all of the linear functions so that their values measure Euclidean distances to the line or plane. Figure 1.7 shows a collection of planes that approximate the handle of a faucet. These planes are the zero contour of a continuous piecewise linear function. In terms of design, these planes may thought of as forming a *control polyhedron* that approximates the shape of some desired C^1 surface.

1.3.2. Approximation of Higher-Order Functions.   Data for free-form blending can also be specified using higher-order functions. For example,

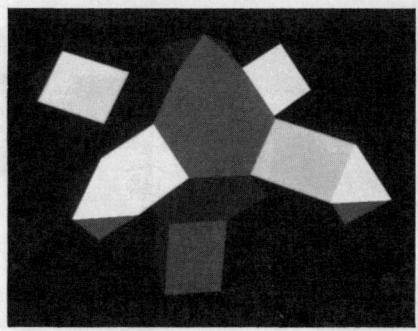

FIG. 1.7.   *A planar approximation to the handle of a faucet.*

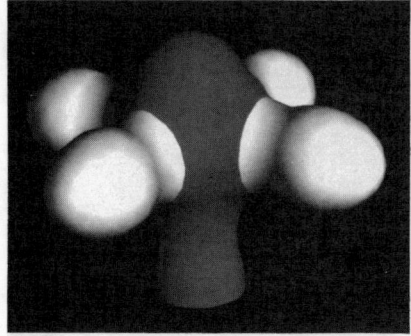

FIG. 1.8. *Faucet after interpolatory blending.*

FIG. 1.9. *Faucet after approximative blending.*

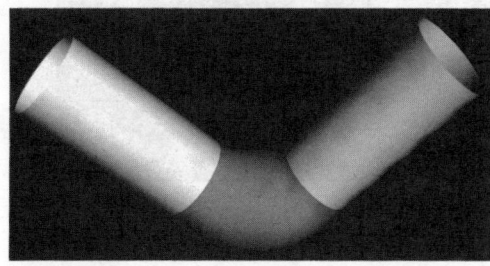

FIG. 1.10.   *A cubic elbow joining two cylinders.*

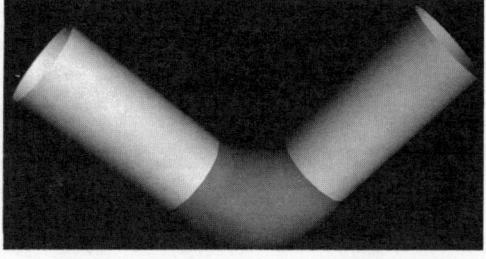

FIG. 1.11. *An approximation to the elbow using free-form blending.*

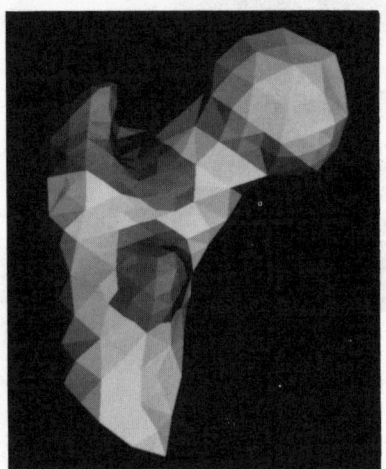

FIG. 1.12. *An adaptive polyhedral model of a femur.*

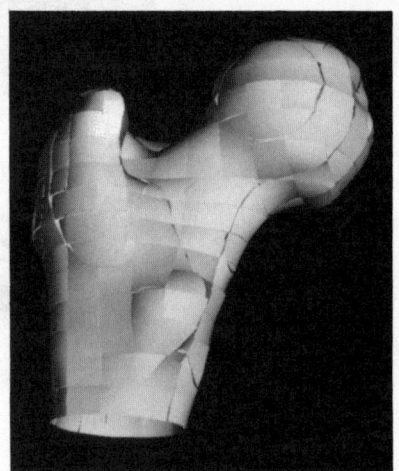

FIG. 1.13. *A discontinuous quadric approximation to the femur.*

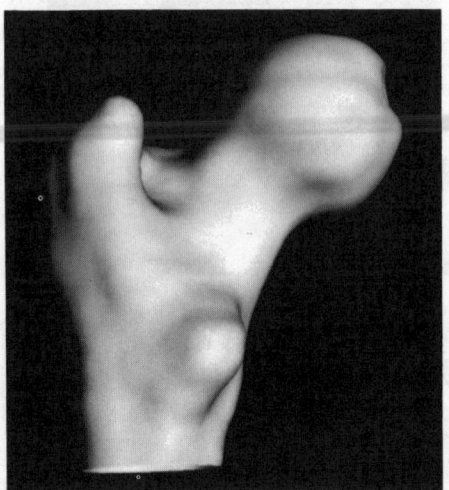

FIG. 1.14. *Quadric femur after interpolatory blending.*

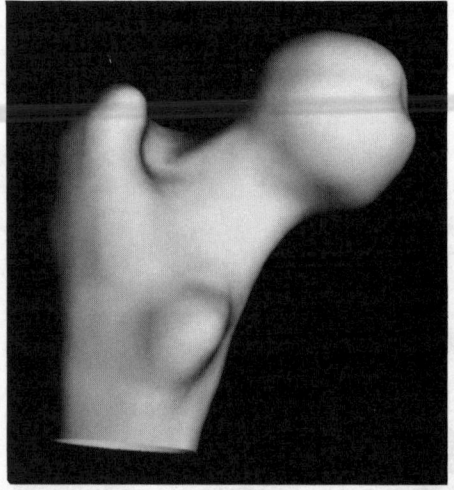

FIG. 1.15. *Quadric femur after approximative blending.*

the designer can specify a collection of quadratic or cubic functions that approximate the desired geometry. Again, the collection of functions need not be continuous over the underlying mesh. Subsequent blending of the C^k data guarantees consistency. Such a strategy allows inclusion of higher-degree surfaces, like those that result from operations such as filleting and blending [16], [22], [7], [11]. For example, two cylinders with equal radii and intersecting axes can be smoothly blended using a single cubic surface [21], as shown in Fig. 1.10. This cubic surface cannot be represented exactly by an interpolating function if the interpolant produces only quadratic surfaces. However, free-form blending leads to a good approximation; we simply extract C^1 data from the quadratic and cubic functions defining the surfaces and construct a piecewise interpolant that strongly resembles the cubic surface.

1.3.3. Scattered Data Approximation.

Free-form blending can also be used to approximate scattered data. In applications such as medical imaging and computer vision, large numbers of data points lying on or near some unknown object can be easily generated. The problem is then to reconstruct the unknown geometry from the data points. The problem of approximating a parametric surface using an implicit surface can easily be recast into a scattered data approximation problem. The parametric functions for the surface can generate as many data points on the parametric surface as necessary. An implicit surface can then be fitted to these data points.

Implicit surfaces appear to be better suited for surface approximation than parametric surfaces because fitting a parametric surface usually requires topological information relating the data points. Several researchers [14], [13] have considered least squares techniques for fitting implicit surfaces to point data. Specifically, Patrikalakis and Kriezis [13] suggest fitting the zero contour of a tensor product B-spline to the scattered data. One difficulty with this approach is that the least squares calculations require time proportional to the square of the number of unknown control points of the B-spline.

Using free-form blending, it is possible to avoid this bottleneck. A simple surface such as a quadric is fit to only those data points located inside one element of the mesh for the interpolant. The fitting process now runs in time proportional to the square of the number of degrees of freedom of the quadric, ten, times the number of elements. C^k data can then be extracted from the patches for use in free-form blending. Another benefit of fitting patches locally is that local estimates of error are possible. If the error exceeds acceptable bounds, then the element can be subdivided and the fitting process repeated for each new subelement.

The author, in collaboration with D. Moore, has implemented a least squares algorithm for use with free-form blending [12]. The method incorporates a new technique for fitting implicit surfaces that avoids undesirable fitting surfaces such those in Fig. 1.1. The technique allows for automatic fitting of scattered data from very complex geometry. Figure 1.13 shows a discontinuous collection of quadric surfaces that approximate the upper portion of a human femur.

1.4. Averaging C^k Data

Using free-form blending, there is no need for a designer to satisfy a complicated set of conditions to ensure various orders of continuity in the final surface. Given the partition of the area of interest into elements, the designer may specify data that are not continuous between adjacent elements. Thus, one vertex in the mesh may have several distinct pieces of C^k data associated with it. To apply a C^k interpolant, these distinct pieces of data must be reconciled to yield one piece of C^k data per vertex in the mesh. The approach suggested here is to average the C^k data. This section discusses some of the averaging techniques that the author has explored for use with free-form blending.

1.4.1. C^0 Averaging.

Free-form blending can be used to create a continuous piecewise linear or multilinear function. Let p be a vertex in the underlying mesh. The designer has specified a function for each element containing p. Evaluating these functions at p yields a set of values $v_1, \cdots, v_k$. To apply linear interpolation to the resulting mesh, these k values must reconciled to a single value at p. The simplest approach is to use a convex combination of the values v_i,

$$v = \sum_{1 \leq i \leq k} \alpha_i v_i,$$

where $\alpha_i \geq 0$, $\sum \alpha_i = 1$. The weights α_i need not be equal; in many instances, the designer or the system may be able to infer that certain data are particularly inaccurate or unreliable. These data may need to be weighted less heavily.

Even though the next section provides practical techniques for creating C^1 surfaces, methods that create C^0 surfaces are still quite important. The simplicity of straight lines and planes makes mathematical analysis of shapes and structures much simpler. For example, in modeling a human femur for surgical planning, a compact, adaptive polyhedral model of the femur is very desirable. This polyhedral model can be used in a finite element analysis of the stress distribution over the femur and kinematic analysis of the joint. Figure 1.12 shows an adaptive polyhedral model that approximates the outer geometry of the femur. The pink portions of the femur indicate areas in which the model was refined into smaller polygons to better model the more complex geometry. This model consists of around 1000 polygons and approximates more than 20,000 data points.

1.4.2. Interpolatory C^1 Averaging.

In the C^1 case, there is not only a value v_i at p from each element containing p, but also derivatives from each element containing p. The simplest strategy is to average the values, as in the C^0 case, and to average the derivatives similarly. An undesirable

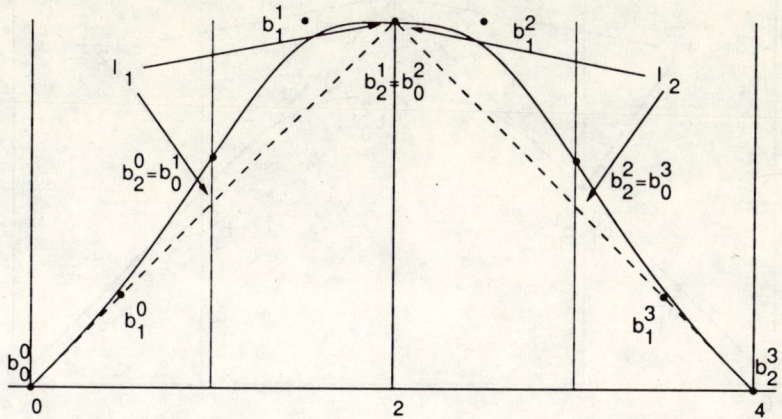

FIG. 1.16. *An interpolating curve can vary more than the data.*

result of this type of averaging is that the resulting functions may vary more than designer's input data do. For example, consider the univariate case in which there are two intervals meeting at a common end point. If the designer specifies a hat-shaped piecewise linear function over these intervals, as in Fig. 1.16, then averaging both the values and derivatives at the middle vertex and applying the piecewise quadratic interpolant of §1.2.2 yields a function whose zero contour passes through the three defining points of the linear function with the proper slopes. However, the curve possesses two points of inflection where the curvature reverses. This phenomenon is quite common in interpolation methods. We call this averaging technique *interpolatory C^1 averaging*.

Some results of interpolatory averaging and the application of triquadratic interpolants appear in Figs. 1.8, 1.11, and 1.14. Averaging the C^1 data from Fig. 1.7 produces data that, when applied to an interpolant, yields the C^1 faucet of Fig. 1.8. Figure 1.11 shows the result when free-form blending is applied to the cubic elbow of Fig. 1.10. The C^1 data were averaged in the interpolatory manner. Figure 1.14 shows the C^1 surface that results from interpolatory blending of the C^1 data from the collection of quadric surface patches in Fig. 1.13.

1.4.3. Approximative C^1 Averaging.

In many applications, the designer wishes to dampen the variation in the data provided rather than increasing it. For curve methods such as Bézier curves and B-spline curves, the variation diminishing property guarantees the desired behavior. As seen in the previous section, interpolatory averaging of C^1 data does not dampen the variation in the original data. However, for cubical meshes, it is possible to average the C^1 data in such a way that the variation in the data is diminished. This averaging technique is general, but described here as it applies specifically to the multiquadratic interpolant of §1.2.2.

In Fig. 1.7, the graph of the interpolant has two inflection points. These inflection points result from the need to interpolate the value of the piecewise

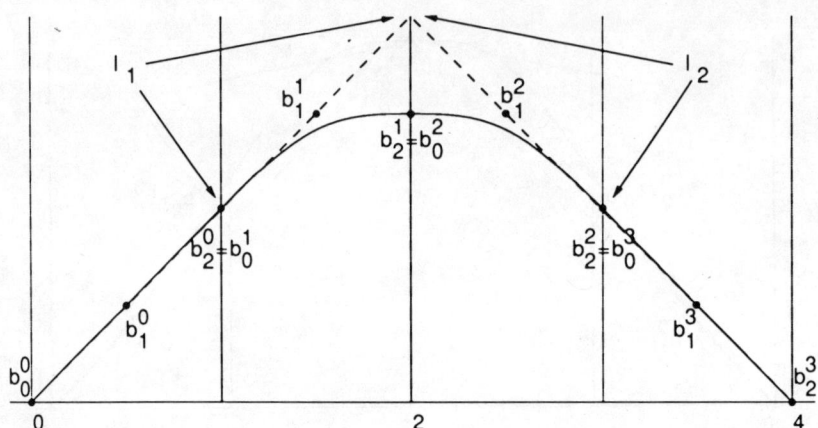

FIG. 1.17. *An approximating curve smoothes the data.*

linear function at the middle vertex. If this restriction is relaxed, then these
inflection points can be avoided. For interpolatory averaging, the coefficients b_1^1
and b_1^2 are chosen based strictly upon the averaged C^1 values at the boundary
point $x = 2$. If b_1^1 and b_1^2 are computed directly from the linear functions l_1
and l_2 used to create the C^1 data at $x = 2$, the resulting piecewise quadratic
curve has no inflection points. If the original C^1 data come from a piecewise
multiquadratic function, b_1^1 and b_1^2 should be exactly the coefficients of the
multiquadratic at the domain points $x = 1.5$ and $x = 2.5$.

Let $l_i(x)$ for $i \in \{1, 2\}$ be the unique linear function defined by the C^1 data
at $x = 2$ associated with the interval $(i, i + 1)$. From the previous discussion,
then, $b_1^1 = l_1(1.5)$ and $b_1^2 = l_2(2.5)$. In order to ensure C^1 continuity in the two
quadratic functions that meet at $x = 2$, set $b_2^1 = b_0^2$ to the average of b_1^1 and b_1^2.
Figure 1.17 illustrates the graph of the piecewise quadratic interpolant applied
to these new C^1 data. Note the new curve has no inflection points.

This averaging can be generalized to n-dimensional cubical meshes in the
following manner. Consider the 2^n n-dimensional hypercubes of size two that
share the common vertex $(2, \cdots, 2)$. Each hypercube has a distinct piece of C^1
data associated with the vertex $(2, \cdots, 2)$. For $i_k \in \{1, 2\}$, $1 \leq k \leq n$, let $l_{i_1 \cdots i_n}$
be the unique linear function defined by the C^1 data at $(2, \cdots, 2)$ associated
with the hypercube that is the product of intervals $\prod_{k=1}^{n}(i_k, i_k + 1)$. Coefficient
$b_{1 \cdots 1}^{i_1 \cdots i_n}$ associated with the domain point $(i_1 + \frac{1}{2}, \cdots, i_n + \frac{1}{2})$ has value

$$b_{1 \cdots 1}^{i_1 \cdots i_n} = l_{i_1 \cdots i_n}\left(i_1 + \frac{1}{2}, \cdots, i_n + \frac{1}{2}\right).$$

The common coefficient at the vertex $(2, \cdots, 2)$ can be chosen to maintain C^1
continuity and be consistent with the neighboring coefficients $b_{1 \cdots 1}^{i_1 \cdots i_n}$ if it is
computed directly using multilinear interpolation over the cube with vertices
whose values are $b_{1 \cdots 1}^{i_1 \cdots i_n}$.

The smoothing effect of this type of averaging relies on the fact that
applying multilinear interpolation to the coefficients $b_{1 \cdots 1}^{i_1 \cdots i_n}$ yields new C^1 data

for $(2, \cdots, 2)$ that is a convex combination of the original coefficients $b_{1 \cdots 1}^{i_1 \cdots i_n}$. This process is called *approximative C^1 averaging*. For simple types of input data, it is possible to show that this type of averaging reduces the fluctuation present in the data.

THEOREM 1.4.1. *Let $f(x_1, \cdots, x_n)$ be a piecewise linear, but possibly discontinuous function defined over 2^n hypercubes that all share a common vertex. Let $g(x_1, \cdots, x_n)$ be the C^1 piecewise multiquadratic function over these hypercubes that results from applying free-form blending using approximative C^1 averaging in the neighborhood of the common vertex. The graph of g lies in the convex hull of the graph of f.*

This theorem directly follows from the fact that the functions $l_{i_1 \cdots i_n}$ in approximative C^1 averaging are exactly the linear functions comprising the piecewise linear function f. The coefficients of the multiquadratric interpolant g are chosen to be convex combinations of points on the graphs of the functions $l_{i_1 \cdots i_n}$. Thus, the graph of the multiquadratic interpolant g must lie in the convex hull of the piecewise linear function f.

If not only first derivatives are available at $(2, \cdots, 2)$, but also all mixed derivatives of order zero or one in each variable, then one possible enhancement to this technique replaces the linear functions $l_{i_1 \cdots i_n}$ by the unique multilinear function $ml_{i_1 \cdots i_n}$ consistent with these mixed derivatives.

Figure 1.9 shows a contour of the trivariate function generated by approximatively blending the C^1 data from Fig. 1.7. Note that this is not the zero contour of the function, but a contour for some larger threshold value. Figure 1.15 shows the C^1 surface that results from approximative blending of the C^1 data from Fig. 1.13. Note that approximative blending decreases the number of variations or "wrinkles" in the surface.

1.5. Further Work

There remain several problems that must be addressed if free-form blending is to be a truly general technique for geometric modeling. A method for performing Boolean set operations on volumes created by free-form blending is essential. In this respect, the mesh on which the interpolant is defined is both a boon and a hindrance. If two volumes share the same mesh, then the subdivision property of the interpolant allows for a simple recursive algorithm for set operations. However, if the volumes do not share a common mesh (say one of the volumes has been rotated), methods for set operations are much trickier.

Another area of importance is the theory of C^1 simplicial interpolants. Current C^1 interpolants in three dimensions are encumbered by fairly complicated geometric constraints. Ideally, free-form blending should be able to be performed on general simplicial meshes. Finally, tighter characterizations of the relationship between the input data and the surface resulting from free-form blending are necessary. Parametric surface techniques always yield surfaces that behave in a well-defined manner. For implicit surfaces to be considered as an alternative to parametric surfaces, such techniques must be developed for implicit surfaces.

Acknowledgment

This research was supported in part by National Science Foundation grants IRI 88-10747 and CCR 89-03431.

References

[1]   P. Alfeld, *A bivariate C^2 Clough-Tocher scheme*, Comput. Aided Geom. Des., 1 (1984), pp. 257–267.

[2]   Chanderjit Bajaj, *Algorithmic implicitization of algebraic curves and surfaces*, Tech. Report CSD-TR-681, Purdue University, Department of Computer Science, 1987.

[3]   Jules Bloomenthal, *Polygonalization of implicit surfaces*, Comput. Aided Geom. Des., 5 (1988), pp. 341–355.

[4]   R. Clough and J. Tocher, *Finite element stiffness matrices for analysis of plates in blending*, in Proceedings of Conference on Matrix Methods in Structural Analysis, 1965.

[5]   Gerald Farin, *Curves and Surfaces for Computer Aided Geometric Design: A Practical Guide*, Academic Press, New York, 1988.

[6]   David R. Forsey and Richard H. Bartels, *Hierarchical B-spline refinement*, Computer Graphics, 22 (1988), pp. 205–212.

[7]   Christoph Hoffmann and John Hopcroft, *The potential method for blending surfaces and corners*, in Geometric Modeling: Algorithms and New Trends, Gerald Farin, ed., Society for Industrial and Applied Mathematics, Philadelphia, 1987.

[8]   K. Höllig, *Box splines*, in Approximation Theory V, C. K. Chui, L. L. Schumaker, and J. D. Ward, eds., Academic Press, New York, 1986.

[9]   ——, *Box-splines surfaces*, in Mathematical Methods in Computer Aided Geometric Design, T. Lyche and L. L. Schumaker, eds., Academic Press, Boston, 1989.

[10] D. Kalra and A. Barr, *Guaranteed ray intersections with implicit surfaces*, Computer Graphics, 23 (1989), pp. 297–306.

[11] A. Middleditch and K. Sears, *Blend surfaces for set theoretic volume modeling systems*, Computer Graphics, 19 (1985), pp. 161–170.

[12] Doug Moore and Joe Warren, *Adaptive approximation of scattered contour data using piecewise implicit surfaces*, Proc. of the Hawaii International Conference on System Sciences, 1990.

[13] Nicholas M. Patrikalakis and George A. Kriezis, *Representation of piecewise continuous algebraic surfaces in terms of B-splines*, The Visual Computer, 5 (1989), pp. 360–374.

[14] Vaughan Pratt, *Direct least-squares fitting of algebraic surfaces*, Computer Graphics, 21 (1987), pp. 145–152.

[15] M. J. D. Powell and M. A. Sabin, *Piecewise quadratic approximations on triangles*, ACM Trans. Math. Software, 3 (1977), pp. 316–325.

[16] Alyn Rockwood and J. Owen, *Blending surfaces in solid geometric modeling*, SIAM Conference on Geometric Modeling and Robotics, Albany, New York, 1985.

[17] J. Rossignac, *Constraints in constructive solid geometry*, 1986 Workshop on Interactive 3D Graphics, Chapel Hill, North Carolina, 1986.

[18] P. Sablonnière, *Composite finite elements of class C^k*, J. Comput. Appl. Math., 12–13 (1985), pp. 542–550.

[19] Thomas W. Sederberg, D. C. Anderson, and Ron Goldman, *Implicit representation of parametric curves and surfaces*, Computer Vision Graphics and Image Processing, 28 (1984), pp. 72–74.

[20] R. Sibson, *A seamed quadratic element for contouring*, Preprint, University of Bath, 1980.

[21] Joe Warren, *Blending quadric surfaces with quadric and cubic surfaces*, in Third ACM Symposium on Computational Geometry, ACM, New York, 1987, pp. 341–347.

[22] ——, *Blending algebraic surfaces*, ACM Trans. Graphics, 8 (1989), pp. 263–278.

[23] T. Whelan, *A representation of a C^2 interpolant over triangles*, Comput. Aided Geom. Des., 3 (1986), pp. 53–66.

[24] A. J. Worsey and Gerald Farin, *An n-dimensional Clough-Tocher element*, Constr. Approx., 3 (1987), pp. 99–110.

[25] A. J. Worsey and B. Piper, *A trivariate Powell-Sabin interpolant*, manuscript, 1989.

Surface Fitting Using Implicit Algebraic Surface Patches

Chandrajit L. Bajaj

2.1. Introduction

Interpolation and least-squares approximation provide efficient ways of generating C^k-continuous meshes of surface patches, necessary for the construction of accurate computer geometric models of solid physical objects (see, e.g., [4], [5]). Two surfaces $f(x, y, z) = 0$ and $g(x, y, z) = 0$ meet with C^k-continuity along a curve C if and only if there exists functions $\alpha(x, y, z)$ and $\beta(x, y, z)$ such that all derivatives up to order k of $\alpha f - \beta g$ equals zero (see [58]). C^k-continuity of two surface patches follows if the above condition is true along the common boundary curves between the two patches.

This paper surveys the use of low degree, implicitly defined, algebraic surfaces and surface patches in three-dimensional real space $\mathbb{R}^3$ for various scattered data fitting problems. The use of low degree algebraic surface patches to construct models of physical objects stems from the advantage of faster computations in subsequent geometric model manipulation operations for design prototyping and manufacturing.

2.1.1. Why Algebraic Surfaces?
A real algebraic surface S in $\mathbb{R}^3$ is implicitly defined by a single polynomial equation $\mathcal{F} : f(x, y, z) = 0$, where coefficients of f are over the real numbers $\mathbb{R}$. Manipulating polynomials, as opposed to arbitrary analytic functions, is computationally more efficient. Furthermore, algebraic surfaces provide enough generality to accurately model most complicated rigid objects.

2.1.2. Why Implicit Representations?
While all real algebraic surfaces have an implicit definition $\mathcal{F}$, only a small subset of these real surfaces can also be defined parametrically by the triple $\mathcal{G}(s, t) : (x = G_1(s, t), y = G_2(s, t), z = G_3(s, t))$ where each G_i, $i = 1, 2, 3$, is a rational function (ratio of polynomials) in s and t over $\mathbb{R}$. The primary advantage of the implicit definition $\mathcal{F}$ is the closure properties of the class of algebraic surfaces under modeling operations such as intersection, convolution, offset, blending, etc. The strictly smaller class of parametrically defined algebraic surfaces $\mathcal{G}(s, t)$ are not closed under

23

any of the operations listed before. Closure under modeling operations allow cascading repetitions[1] without any need of approximation. Furthermore, designing with a larger class of surfaces leads to better possibilities (as we show here) of being able to satisfy the same geometric design constraints with much lower degree algebraic surfaces. The implicit representation of algebraic surfaces also naturally yields sign-invariant regions $\mathcal{F}^+ : f(x,y,z) \geq 0$ and $\mathcal{F}^- : f(x,y,z) \leq 0$, a fact quite useful for intersection and offset modeling operations. Finally, prior approaches to interpolation and least-squares fitting to scattered data in three dimensions have focused primarily on the parametric representation of surfaces [19], [23], [31]–[33], [47], [54], [56], [60]. Our aim here is to exhibit that implicitly defined algebraic surfaces are also very appropriate for geometric surface design.

2.1.3. Additional Notation and Definitions.

A real algebraic space curve can be implicitly defined as the common intersection of two or more real algebraic surfaces $C : (f_1(x,y,z) = 0, f_2(x,y,z) = 0, f_3(x,y,z) = 0, \cdots)$. A smaller class of *rational* algebraic space curves can also be represented by the triple $\mathcal{H}(s) : (x = H_1(s), y = H_2(s), z = H_3(s))$, where H_1, H_2, and H_3 are rational functions in s over $\mathbb{R}$. Whenever we consider the special case of a rational space curve, we assume that the curve is smooth and only singly defined under the parameterization map, i.e., a triple of values for $(x,\ y,\ z)$ corresponds to a single value of s and this is true for all but a finite number of points on the curve.

The "normal" N_p of a point p is an arbitrary nonzero vector associated with p. N_p defines a unique plane containing p. The "normal" N_C of a curve C is a one-dimensional set of vectors, one vector associated with each point p on C, and orthogonal to the tangent vector at p. We assume the input curves are smooth, i.e., nonsingular, though this is not a necessary requirement. Finally, a surface patch is defined as a smooth, connected two-dimensional region of a surface bounded by a single cycle of curve segments.

2.1.4. Problem Descriptions.

1. C^k-*Interpolation Surface Fit*: Construct a single real algebraic surface S which C^k-interpolates a collection of l points $\mathbf{p}_i$ in $\mathbb{R}^3$ with associated fixed "normal" unit vectors $\mathbf{m}_i$, and m given space curves C_j in $\mathbb{R}^3$, possibly with associated "normal" unit vectors $\mathbf{n}_j$ and additionally up to kth-order derivatives[2] of $\mathbf{n}_j$, varying along the entire span of the curves. Assume that any of the vectors $\mathbf{m}_i$ and $\mathbf{n}_j$ or their derivatives are never identically zero, a phenomenon that occurs at singularities. By C^k-interpolation we shall mean that the interpolating surface S contains each of the points and curves and furthermore has its gradient together

---

[1] The output of one operation acts as the input to another operation.

[2] The emphasis being algebraic space curves, the "normals" and higher order derivatives along curves are restricted to polynomials of some degree.

with its $1 \cdots k$th-order derivatives, respectively, in the same direction as the specified "normal" vectors and its derivatives along the entire span of the C_js. [*This is one natural generalization into space of the usual two-dimensional Hermite interpolation, applied to fitting curves through point data and matching derivatives at those points.*]

2. C^k *Least-Squares Approximate Surface Fit*: Construct a real algebraic surface S, which C^{k-1}-interpolates a collection of points $\mathbf{p}_i$ in $\mathbb{R}^3$ and given space curves C_j in $\mathbb{R}^3$ as before, with associated unit "normal" vectors and its $1 \cdots (k-1)$th-order derivatives, and additionally minimizes the Euclidean 2-norm of the difference of the kth-order derivative of Ss normalized gradient and the kth-order derivative of the specified unit "normal" vectors, on the same collection of points and space curves. [*This is a natural generalization of ordinary C^0 least-squares approximation (the case $k = 0$) which minimizes only the sum of the squares of the distances of the solution from a collection of points or curves.*]

3. C^k-*Interpolation and C^l Least-Squares Surface Patches*: Construct a mesh of real algebraic surface patches S_i, which C^k-interpolates a collection of points $\mathbf{p}_i$ in $\mathbb{R}^3$ and given space curves C_j in $\mathbb{R}^3$, with associated "normal" unit vectors and their derivatives, varying along the entire span of the curves and C^l least-squares approximates a collection of points $\mathbf{q}_i$ in $\mathbb{R}^3$ and given space curves D_j in $\mathbb{R}^3$ with associated "normal" unit vectors and their derivatives, varying along the entire span of the curves.

4. *Triangulated Data Fit with Surface Patches*: Given a triangulation $\mathcal{T}$ of points $\mathbf{p} = (x_i, y_i, z_i)$ in $\mathbb{R}^3$, possibly with various order derivatives at these points, construct a C^k-continuous mesh of real algebraic surface patches S_i, respecting the topology of the triangulation $\mathcal{T}$, and which C^k-interpolates the collection of points $\mathbf{p}_j = (x_j, y_j, z_j)$ in $\mathbb{R}^3$ and derivatives at those points. Additionally, this C^k mesh of patches may also C^l least-squares approximate a collection of points $\mathbf{q}_k = (x_k, y_k, z_k)$ and derivatives in $\mathbb{R}^3$.

5. *Interactive Shape Control of Implicit Surface Families*: Interactively control the shape of an interpolating or approximating implicit surface by selecting appropriate instances from a p-parameter family of solution surfaces. [*Such p-parameter families of surfaces are the result of the above C^k-interpolatory fits.*]

2.1.5. Paper Outline. The rest of the paper is structured as follows. Each of the subsequent §§2.2–2.6 is devoted to one of the above problems and summarizes various recent approaches to implicit surface fitting for the appropriate problem. The section then details a recent result, which the author is most familiar with, and provides examples to clarify the algorithm presented.

2.2. C^k-Interpolation Surface Fit

2.2.1. Problem. Construct a single real algebraic surface S which C^k-interpolates a collection of l points $\mathbf{p}_i$ in $\mathbb{R}^3$ with associated fixed "normal" unit vectors $\mathbf{m}_i$, and m given space curves C_j in $\mathbb{R}^3$, possibly with associated "normal" unit vectors $\mathbf{n}_j$ and additionally up to kth-order derivatives of $\mathbf{n}_j$ varying along the entire span of the curves.

2.2.2. Summary of Approaches. There has been extensive prior work in interpolatory or exact surface fitting through scattered data. Much of it has either concentrated on polynomial parametric (and occasionally rational parametric) surface fitting through scattered point data in three dimensions (see, e.g., the surveys by Alfeld [2], Boehm et al. [18], Franke [26], Sabin [52]). Exact fitting of curves (primarily conics) has been considered by several authors (see, e.g., [14], [17], [29], [40], [46], [53]. An exposition of exact C^0 fitting of implicitly defined algebraic surfaces through given data points is presented in [50]. Characterizations of C^0 surface fits of points and curves using implicitly defined algebraic surfaces is also given by [55]. Furthermore, implicit algebraic surfaces have been used for C^1 blend fits [19], [41], [45], [51], [58]. Other approaches to parametric surface fitting and transfinite interpolation are also mentioned in that paper, as well as in [25], [47], [44], [56], [60]. Papers [11], [12] generalize the results of [50], [55]. It provides conditions for exact C^k fits of implicitly defined algebraic surfaces through given points and space curves together with derivative information ("normals") along the curves.

2.2.3. Recent Results. Bajaj and Ihm [9] present a simple constructive characterization of the real algebraic surface which C^1-interpolates any given number of points and algebraic space curves, with associated "normal" directions. This characterization, called *Hermite interpolation*, deals with the containment and matching normals at the points or varying along the entire span of the space curves. The input for Hermite interpolation is a description of the properties of a surface to be designed in terms of a combination of points and curves, possibly associated with "normal" directions. For an algebraic surface S of degree n, C^1-interpolation generates a homogeneous linear system $\mathbf{M_I}\,\mathbf{x} = \mathbf{0}$ where $\mathbf{x}$ is a $\binom{n+3}{3}$-vector[3] of the coefficients of the algebraic surface S. All nontrivial vectors, if any, in the nullspace of $\mathbf{M_I}$ form a family of all the surfaces, satisfying the given description. The coefficients of the family of surfaces are expressed in terms of p-parameters where p is the rank of the nullspace.

In C^1-interpolation, smoothness is achieved by making the normals of tangent planes of the surface to be designed identical to those of given points

---

[3]There are $\binom{n+3}{3}$ coefficients in $f(x, y, z)$ of degree n.

or curves. For some applications of modeling, such as design of the body of an airplane, however, more than tangent plane smoothness is desirable. This concept of smoothness is generalized by defining a higher order of geometric continuity. DeRose [23] gives such a definition between parametric surfaces, where two surfaces F_1 and F_2 meet with order k geometric continuity (concisely stated as $\mathbf{C}^k$-continuity) along a curve C if and only if there exist local reparameterizations F_1' and F_2' of F_1 and F_2, respectively, such that all partial derivatives of F_1' and F_2' up to degree k agree along C. Warren [58] formulates an intuitive definition of $\mathbf{C}^k$-continuity between implicit surfaces as follows.

DEFINITION 2.2.1. *Two algebraic surfaces $f(x, y, z) = 0$ and $g(x, y, z) = 0$ meet with $\mathbf{C}^k$-continuity at a point p or along an irreducible algebraic curve C if and only if there exists two polynomials $a(x, y, z)$ and $b(x, y, z)$, not identically zero at p or along C, such that all derivatives of $a \cdot f - b \cdot g$ up to degree k vanish at p or along C.*

This formulation is *more general* than just making all the partials of $f(x, y, z) = 0$ and $g(x, y, z) = 0$ agree at a point or along a curve. For example, consider the intersection of the cone $f(x, y, z) = xy - (x + y - z)^2 = 0$ and the plane $g(x, y, z) = x = 0$ along the line defined by two planes $x = 0$ and $y = z$. It is not hard to see that these two surfaces meet smoothly along the line since the normals to $f(x, y, z) = 0$ at each point on the line are scalar multiples of those to $g(x, y, z) = 0$. But, this scale factor is a function of z. Situations like these are corrected by allowing multiplication by certain polynomials, not identically zero along a intersection curve. Note that multiplication of a surface by polynomials nonzero along a curve does not change the geometry of the surface in the neighborhood of the curve. Garrity and Warren in [28] also prove that this notion of rescaling C^k-continuity is equivalent to other kth-order derivative continuity measures as well as to reparameterization continuity for parametric surfaces. In [11], Bajaj and Ihm show how to form a C^1-interpolation matrix $\mathbf{M_I}$ and proved that using this one is able to construct all surfaces meeting each other with rescaling $\mathbf{C}^1$-continuity. However, even though one is currently unable to translate geometric specifications for $\mathbf{C}^k$-continuity ($k \geq 2$) into a matrix $\mathbf{M_I}$ whose nullspace captures *all* $\mathbf{C}^k$ continuous surfaces, from the theorem below one can generate an interpolation matrix $\mathbf{M_I}$ whose nullspace captures an interesting proper subset of the whole class.

THEOREM 2.2.1 ([12]). *If surfaces $g(x, y, z) = 0$ and $h(x, y, z) = 0$ intersect transversally along an irreducible curve C, then any algebraic surface $f(x, y, z) = 0$ that meets $g(x, y, z) = 0$ with C^k-continuity along C must be of the form $f(x, y, z) = \alpha(x, y, z)g(x, y, z) + \beta(x, y, z)h^{k+1}(x, y, z)$. If $g(x, y, z) = 0$ and $h(x, y, z) = 0$ share no common components at infinity, then the degree of $\alpha(x, y, z)g(x, y, z) \leq$ degree of $f(x, y, z)$ and the degree of $\beta(x, y, z)h^{k+1}(x, y, z) \leq$ degree of $f(x, y, z)$.*

For given curves C_i, $i = 1 \cdots l$, which are, respectively, the transversal intersection of given surfaces $g_i(x, y, z) = 0$ and $h_i(x, y, z) = 0$, a surface $f(x, y, z) = 0$ containing space curves C_i with C^k-continuity then can be constructively obtained by the relations for $i = 1 \cdots l$,

$$(2.1) \qquad f(x,y,z) = \alpha_i(x,y,z)g_i(x,y,z) + \beta_i(x,y,z)h_i^{k+1}(x,y,z).$$

Since the g_i and h_i are known surfaces, the unknown coefficients are those of f, α, and β. Note from the above theorem for a possible interpolating surface f of degree n, both polynomials α and β are of bounded degree. From the relations in (2.1) one sees that these unknown coefficients form a system of linear equations, yielding the interpolation matrix $\mathbf{M_I}$. For the special case of parametric space curves with parametric "normal" and derivative information, the above technique can also be adapted to provide C^k-continuous algebraic surface fits. Here using C^1-interpolation [11], implicit surfaces are first constructed which contain the parametric curve as well as have matching "normals" and derivative information. These implicit surfaces are then used above, to generate matrix $\mathbf{M_I}$ for C^k-continuous fits.

2.2.4. Examples.

Example 2.2.1. *A Quartic Surface for a C^1 Blend of a Corner.* The edges of the table corner are given by $C_1 : (y^2 + z^2 - 25 = 0, x = 0)$, and $C_2 : (x^2 + z^2 - 25 = 0, y = 0)$. Each curve is associated with a "normal" direction which is chosen in the same direction as the gradients of the side of table, the cylinder in C_1 and C_2. That is, $\mathbf{n_1}(x,y,z) = (0, 2y, 2z)$, and $\mathbf{n_2}(x,y,z) = (2x, 0, 2z)$.

The interpolation matrix $\mathbf{M_I}$ is of size 32×35 (32 linear equations and 35 coefficients for a quartic surface) whose rank is 24. The nullspace of $\mathbf{M_I}$ is of dimension 11 represented by a family of quartic surfaces which blend the corner $f(x,y,z) = r_1 z^4 + (r_2 y + r_6 x + 5r_4)z^3 + (r_3 y^2 + (r_7 x + 5r_8)y + r_{10}x^2 + 5r_{11}x - 25r_9 - 25r_1)z^2 + (r_2 y^3 + (r_6 x + 5r_4)y^2 + (r_2 x^2 - 25r_2)y + r_6 x^3 + 5r_4 x^2 - 25r_6 x - 125r_4)z + (r_3 - r_1)y^4 + (r_7 x + 5r_8)y^3 + (r_5 x^2 + 5r_{11}x - 25r_9 - 25r_3 + 25r_1)y^2 + (r_7 x^3 + 5r_8 x^2 - 25r_7 x - 125r_8)y + (r_{10} - r_1)x^4 + 5r_{11}x^3 + (-25r_9 - 25r_{10} + 25r_1)x^2 - 125r_{11}x + 625r_9.$

An instance $f(x,y,z) = -1250 - x^4 - y^4 - x^2 z^2 - y^2 z^2 + 50z^2 + 75y^2 + 75x^2$ of this family is shown with the table in Fig. 2.1. □

Example 2.2.2. *A Quartic Interpolating Surface for a C^1 Join of Four Parallel Cylindrical Surfaces.* In this example, the lowest degree surface is constructed, which smoothly joins four truncated parallel circular cylinders defined by $CYL_1 : y^2 + z^2 - 1 = 0$ for $x \geq 2$, $CYL_2 : y^2 + z^2 - 1 = 0$ for $x \leq -2$, $CYL_3 : (y-4)^2 + z^2 - 1 = 0$ for $x \geq 2$, and $CYL_4 : (y-4)^2 + z^2 - 1 = 0$ for $x \leq -2$.

The C^1-interpolation technique shows that the minimum degree for such joining surface is 4, and finds a two-parameter (one independent parameter) family of algebraic surfaces which is $f(x,y,z) = \frac{r_1}{14}z^4 + \frac{r_1}{7}y^2 z^2 - \frac{4r_1}{7}yz^2 + r_1 z^2 + \frac{r_1}{14}y^4 - \frac{4r_1}{7}y^3 + r_1 y^2 + \frac{4}{7}r_1 y + \frac{14r_2 + 15r_1}{224}x^4 - \frac{14r_2 + 15r_1}{28}x^2 + r_2.$

An instance of this family ($r_1 = 392$, $r_2 = -868$) is shown in Fig. 2.2.

Example 2.2.3. *Cubic and Quartic Surfaces Interpolating with C^2- and C^3-Continuity.* Consider a space curve C defined by the two equations

FIG. 2.1.   *Corner blending with a quartic surface.*

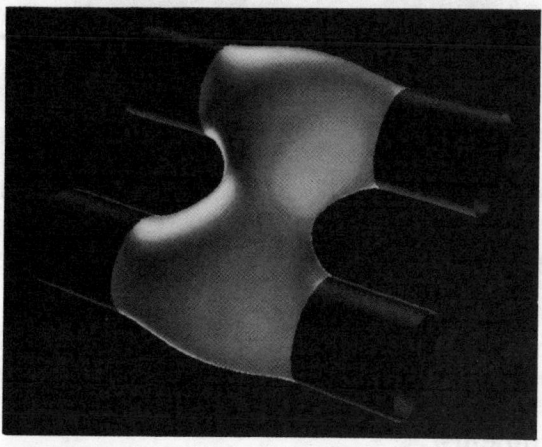

FIG. 2.2. C^1 *join of four cylinders with a quartic surface.*

$f_1(x,y,z) = x^2 + 2y^2 + 2z^2 - 2 = 0$ and $f_2(x,y,z) = x = 0$. A cubic surface C_1 is constructed which interpolates C with C^2 continuity as follows. The general implicit equation of a cubic algebraic surface is given by $f_3(x,y,z) = ax^3 + by^3 + cz^3 + dx^2y + exy^2 + fx^2z + gxz^2 + hy^2z + iyz^2 + jxyz + kx^2 + ly^2 + mz^2 + nxy + oyz + pxz + qx + ry + sz + t = 0$. Using relation (2.1) for C^2-continuity as given in §2.2, one obtains $f_3(x,y,z) = (r_1x + r_2y + r_3z + r_4)f_1(x,y,z) + r_5f_2(x,y,z)^3$ yielding the system of linear equations $a - r_1 - r5 = 0$, $b - 2r_2 = 0$, $c - 2r_3 = 0$, $d - r2 = 0$, $e - 2r_1 = 0$, $f - r_3 = 0$, $g - 2r_1 = 0$, $h - 2r_3 = 0$, $i - 2r_2 = 0$, $j = 0$, $k - r_4 = 0$, $l - 2r_4 = 0$, $m - 2r_4 = 0$, $n = o = p = 0$, $q + 2r_1 = 0$, $r + 2r_2 = 0$, $s + 2r_3 = 0$, $t + 2r_4 = 0$ in unknowns $a, \cdots, t$ and $r_1, \cdots, r_5$. For $r_1 = 1$, $r_2 = -1$, $r_3 = 1$, $r_4 = 1$, $r_5 = 2$, the cubic surface $f_3(x,y,z) = 2z^3 - 2yz^2 + 2xz^2 + 2z^2 + 2y^2z + x^2z - 2z - 2y^3 + 2xy^2 + 2y^2 - x^2y + 2y + 3x^3 + x^2 - 2x - 2$ is shown in Fig. 2.3.

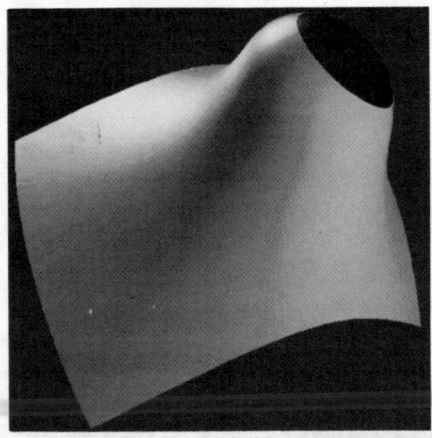

FIG. 2.3.   *Example of C^2-continuous surfaces.*

In the same way, a quartic surface $f_4(x,y,z) = 16z^4 - 16yz^3 + 32xz^3 + 32z^3 + 16y^2z^2 - 16xyz^2 - 16yz^2 + 24x^2z^2 + 32xz^2 - 16y^3z + 32xy^2z + 32y^2z - 8x^2yz + 16yz + 32x^3z + 16x^2z - 32xz - 32z - 9y^4 - 16xy^3 - 16y^3 + 16x^2y^2 + 32xy^2 + 16y^2 - 8x^3y - 8x^2y + 16xy + 16y + 24x^4 + 32x^3 - 8x^2 - 32x - 16$ is constructed which meets f_3 with C^3-continuity along the curve defined by f_3 and $f_5(x,y,z) = y = 0$ as shown in Fig. 2.4. $\square$

2.2.5. Open Problems.

1. Reduce implicit surface interpolation for higher geometric continuity to a linear system that captures all possible solutions?

2. Investigate the relationship of the degrees and relative topology of the input curves with the rank of the interpolation matrices.

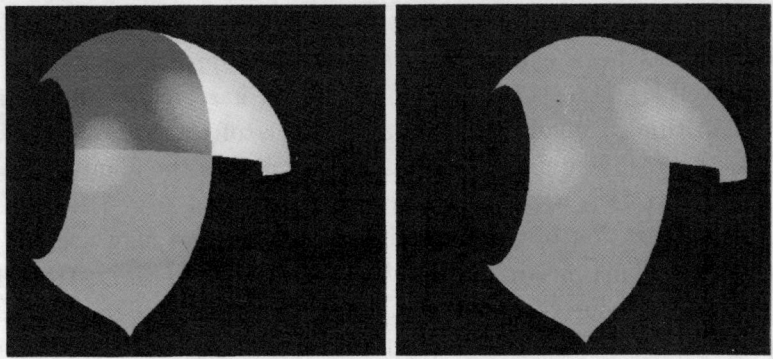

FIG. 2.4. *Example of C^3-continuous surfaces.*

2.3. C^k Least-Squares Approximate Surface Fit

2.3.1. Problem. Construct a real algebraic surface S, which C^{k-1}-interpolates a collection of points $\mathbf{p}_i$ in $\mathbb{R}^3$ and given space curves C_j in $\mathbb{R}^3$ as before, with associated unit "normal" vectors and its $1\cdots(k-1)$th-order derivatives, and additionally minimizes the Euclidean two-norm of the difference of the kth-order derivative of S's gradient and the kth-order derivative of the specified unit "normal" vectors, on the same collection of points and space curves.

2.3.2. Summary of Approaches. The concept of C^k least-squares fitting and also through a mixture of point and curve three-dimensional data is surprisingly novel and a search of the past literature failed to reveal a suitable reference. Pratt [50] and some others [37] consider the traditional C^0 least-squares approximation problem using implicit algebraic surface for only scattered point data. In Bajaj, Ihm, and Warren [12], a C^k-interpolating/least-squares approximating implicit algebraic surface is found by solving a quadratic optimization problem constructed from given sets of three-dimensional points and curves data. In this method, higher order derivative information of points and space curves as well as positional data is interpolated and approximated to. For example, when a surface of some fixed degree does not have sufficient flexibility to C^1-interpolate a set of curves with normal directions, the tangential constraints are least-squares approximated after the positional constraints are exactly C^0-interpolated.

2.3.3. Recent Results. In C^k-interpolation of the previous subsection, one seeks a nontrivial solution $\mathbf{x}$ which is the coefficient vector of an algebraic surface $f(x, y, z) = 0$. To use least-squares approximations for geometric design, one needs to define distance metrics that are meaningful and also computationally viable. As $f(x, y, z) = cf(x, y, z)$ for $c \neq 0$, the coefficients of f are first normalized so that $f(x, y, z)$ represents the equivalence class $\{cf(x, y, z) | c \neq 0\}$. There are infinite number of ways to normalize $f(x, y, z) =$

0. Bajaj, Ihm, and Warren [12] choose to adopt *quadratic normalization*. This normalization has been extensively used in the approximate fitting of conic sections as well [1], [14], [17], [29], [46], [53] and has yielded computationally efficient algorithms. Quadratic normalization is of the form $\mathbf{x}^T \mathbf{M_N} \mathbf{x} = 1$ where $\mathbf{M_N}$ is a real symmetric matrix. In most cases the identity matrix $\mathbf{I}$ or a diagonal matrix $\mathbf{D}$ is used for the matrix $\mathbf{M_N}$.

Once normalization of the coefficients is done, one may use $\|f(p)\|$ as a distance metric. This metric, called *the algebraic distance*, is straightforward to compute and in some cases closely approximates the real geometric distance (the Euclidean distance between a point and a surface). Sampson [53] proposes the use $\|\frac{f(p)}{\nabla f(p)}\|$ as a distance measure (a nonalgebraic distance). Perhaps a better approximation is achievable in some cases, however, only at the enormous cost of iterative applications of least-squares approximation.

Least-squares approximation can be directly used to control the geometric shape of a desired solution interpolating surface. When the rank r of $\mathbf{M_I}$ of §2.3 is less than $\binom{n+3}{3}$, the number of the unknown surface coefficients, there exists a family $f(x, y, z)$ of algebraic surfaces that satisfy the given geometric constraints and whose coefficients are expressed in terms of $p = \binom{n+3}{3} - r$ parameters. The problem of interactively selecting an instance from the solution family is addressed in §2.6. Selecting an instance from the family is equivalent to assigning values to each of the p parameters. When there are p parameters to be instantiated, one may additionally specify a set of points, curves, or even surfaces around the earlier given input data, which approximately describes the final surface to be designed. The final solution instance is computed via interpolation of the given input data and with least-squares approximation of the additional data set. In all these cases, from the matrix $\mathbf{M_I}$ one is easily able to construct a matrix $\mathbf{M_A}$ under the appropriate normalization $\mathbf{M_N}$, such that $\|\mathbf{M_A} \mathbf{x}\|^2 = \mathbf{x}^T \mathbf{M_A}^T \mathbf{M_A} \mathbf{x}$ is minimized. The normalization eliminates certain columns of the matrix $\mathbf{M_I}$ yielding an overdetermined reduced matrix $\mathbf{M_A}$ in the standard way [37].

Bajaj, Ihm, and Warren [12] provide an efficient algorithm based on the orthogonal decomposition of the matrix and computation of eigenvalues and eigenvectors. As a means of computing the nullspace of the system, one uses the *QR method* based upon the Householder's transformation [30], [37]. In order to correctly decide the rank of the matrix during the Householder's transformation, the elements of a lower right part of a matrix are checked for zeros at each step.

2.3.4. Examples.

Example 2.3.1. *A Quartic Surface with C^1-Interpolation of Curves and C^0 Least-Squares Approximation from an Additional Set of Points.* In this example, a quartic surface $f(x, y, z) = 0$ C^1 interpolates curves on the four cylinders described by $CYL_1 : y^2 + z^2 - 1 = 0$ for $x \geq 2$, $CYL_2 : y^2 + z^2 - 1 = 0$ for $x \leq -2$, $CYL_3 : x^2 + y^2 - 1 = 0$ for $z \geq 2$, and $CYL_4 : x^2 + y^2 - 1 = 0$

for $z \leq -2$. As a byproduct during interpolation, it is found that degree 4 is
the minimum required. For this interpolation, $\mathbf{M_I}$ is of size 64×35 (64 linear
equations and 35 coefficients with rank 33), yielding a two-parameter family of
quartic surfaces satisfying the C^1-interpolation constraints. A specific member
of the two-parameter family of surfaces is next selected, via C^0 least-squares
approximation from a collection of new data points. For the normalization
$\mathbf{M_N}$, an identity matrix $\mathbf{I}_{35}$ is used. To illustrate the shaping effect of the
approximation, two independent sets of data points are used: $S_1 = \{(0, 1.75, 0),$
$(0, -1.75, 0), \ (-1, 1.25, 0), \ (-1, -1.25, 0), \ (1, 1.25, 0), \ (1, -1.25, 0)\}$ and $S_2 =$
$\{(0, 1.25, 0), \ (0, -1.25, 0), \ (-0.5, 1.125, 0), \ (-0.5, -1.125, 0), \ (0.5, 1.125, 0)4,$
$(0.5, -1.125, 0)\}$. (See Fig. 2.5(a)(b)).

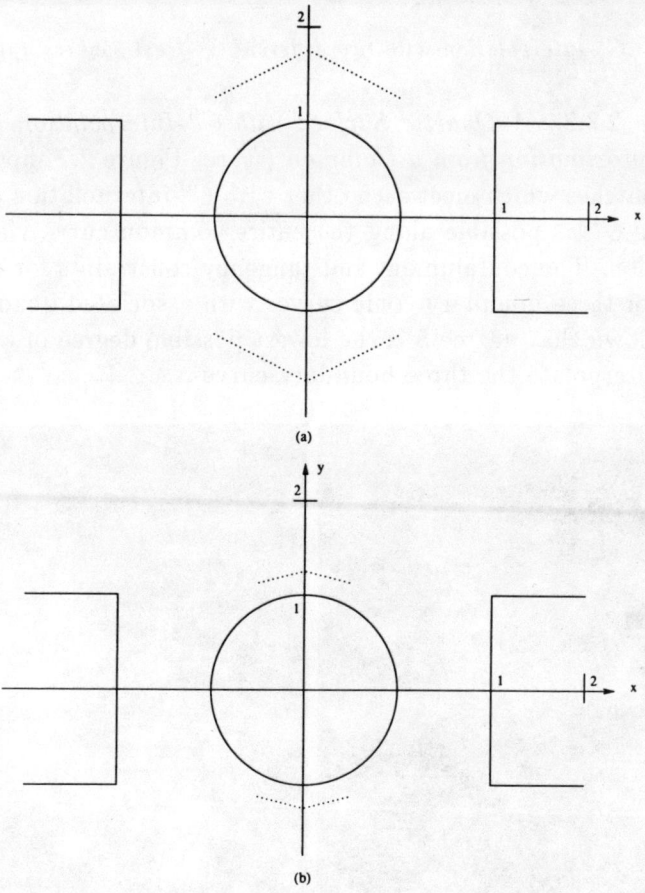

(a)

(b)

FIG. 2.5. *Additional data points for shape control using weighted least-squares
approximation.* (a) *Six points in* S_1. (b) *Six points in* S_2.

For the least-squares approximation with normalization, the eigenvalues
and eigenvectors for S_1 and S_2 are computed. As a result, one obtains
$\lambda_{min_{S_1}} = 1.2546390$, $\lambda_{min_{S_2}} = 0.6439209$, $\mathbf{y}_{min_{S_1}} = [-0.1111540, 0.9938032]^t$,

and $\mathbf{y}_{min_{S_2}} = [0.01853292, 0.9998283]^t$. The corresponding surfaces after normalization are shown in Fig. 2.6(a)(b).   □

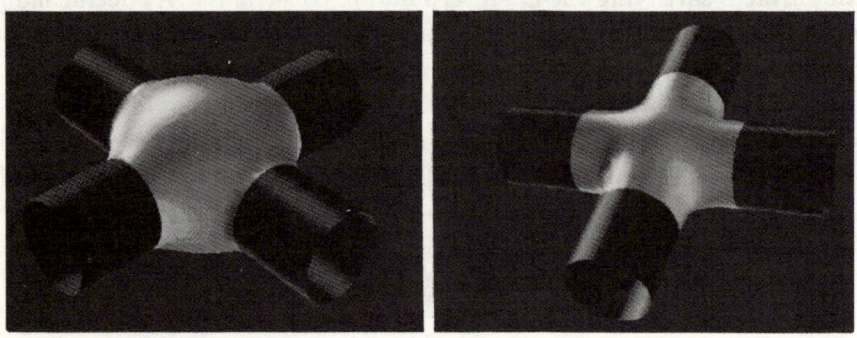

FIG. 2.6. *C^1-interpolation with two different C^0 least-squares approximations.*

Example 2.3.2. A Quartic Surface with C^0-Interpolation and C^1 Least-Squares Approximation from a Common Curve. Figure 2.7 shows two quartic triangluar patches which meet each other with C^0-interpolation continuity and are made as C^1 as possible along the entire common curve via least-squares approximation. The containment and tangency constraints for each patch are generated for three boundary conic curves with associated quadratic normals. It can be shown that degree 5 is the lowest possible degree of a surface which would C^1-interpolate the three boundary curves.   □

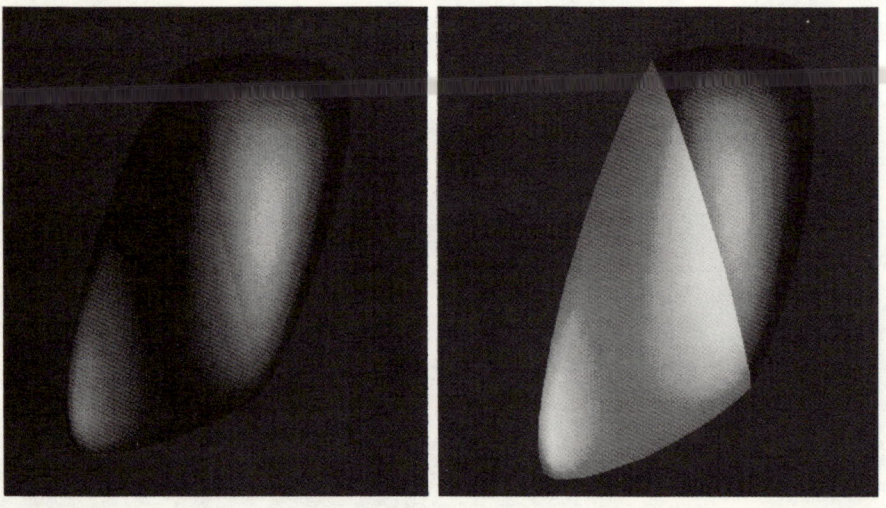

FIG. 2.7. *C^0-interpolation with C^1 least-squares approximation.*

2.3.5. Open Problems.

1. Produce well-conditioned matrices $\mathbf{M_I}$ and $\mathbf{M_A}$ for interpolation and approximation by appropriate choice of basis for the implicit surfaces.

2. Investigate the relationship between stability of computation and topology of the geometric data .

2.4. C^k-Interpolation and C^l Least-Squares Fit with Surface Patches

2.4.1.   Problem.   Construct a mesh of real algebraic surface patches S_i, which C^k-interpolates a collection of points $\mathbf{p}_i$ in $\mathbb{R}^3$ and given space curves C_j in $\mathbb{R}^3$, with associated "normal" unit vectors and their derivatives, varying along the entire span of the curves and C^l least-squares approximates a collection of points $\mathbf{q}_i$ in $\mathbb{R}^3$ and given space curves D_j in $\mathbb{R}^3$ with associated "normal" unit vectors and their derivatives, varying along the entire span of the curves.

2.4.2.   Summary of Approaches.   Solving a linear system of equations plays a key role in C^k-interpolation and approximation of the previous two sections. This section presents another approach of algebraic surface design [6], where a *nonlinear* system of polynomial equations needs to be solved. The emphasis here is on constructing C^k-continuous meshes of implicit surface patches. Such "smooth" meshing has been largely addressed by [34], [47], [54] amongst others, using the Bézier representations of functional or parametric surfaces.

2.4.3.   Recent Results.   The techniques of [6] are primarily based on Bezout's intersection theorem (see [27], [61]).

THEOREM 2.4.1. *If an algebraic surface S of degree n intersects an algebraic surface T of degree m in a curve of degree d with intersection multiplicity i, then $i * d \leq nm$ and a theorem from [12],*

THEOREM 2.4.2. *If surfaces $f(x,y,z) = 0$ and $g(x,y,z) = 0$ intersect transversally in a single irreducible curve[4] C, then any algebraic surface $h(x,y,z) = 0$ contains C with C^k-continuity must be of the form $h(x,y,z) = \alpha(x,y,z)f(x,y,z) + \beta(x,y,z)g^{k+1}(x,y,z)$. Furthermore, the degree of $\alpha(x,y,z)f(x,y,z) \leq$ degree of $h(x,y,z)$ and the degree of $\beta(x,y,z)g^{k+1}(x,y,z) \leq$ degree of $h(x,y,z)$.*

Another required theorem relates continuity with the intersection multiplicity of smooth algebraic surfaces (see [27], [28]).

THEOREM 2.4.3. *Two smooth algebraic surfaces $S_1 : f(x,y,z) = 0$ and $S_2 : g(x,y,z) = 0$ meet with C^k-continuity along a curve C if and only if S_1 and S_2 intersect with multiplicity $k + 1$ along C.*

From Theorem 2.4.3 one obtains the following special case lemma.

LEMMA 2.4.1. *Let $S : f(x,y,z) = 0$ be an irreducible quadric surface, and $Q : q(x,y,z) = 0$ be a plane that intersects S in a conic C. Then, another quadric surface $S_1 : f_1(x,y,z)$ is tangent to S along C if and only if there exists nonzero constants α, β (possibly complex) such that $f_1 = \alpha f + \beta q^2$.*

---

[4]More precisely surfaces $f(x,y,z) = 0$ and $g(x,y,z) = 0$ intersect properly and share no common components at infinity.

Since one is interested in surface fitting with real surfaces, α and β are restricted to real numbers. A related theorem can be derived for the quadric surface interpolation of two conics in space.

LEMMA 2.4.2. *Consider quadrics* $S_1 : f_1 = 0$, $S_2 : f_2 = 0$ *and planes* $Q_1 : q_1 = 0$, $Q_2 : q_2 = 0$. *Let* $C_1 : (f_1 = 0, q_1 = 0)$ *and* $C_2 : (f_2 = 0, q_2 = 0)$ *be two conics in space. Then* C_1 *and* C_2 *can be Hermite interpolated by a quadric surface* S *if and only if there exist nonzero constants* α_1, α_2, β_1, *and* β_2 *(possibly complex) such that* $\alpha_1 f_1 + \beta_1 q_1^2 - \alpha_2 f_2 - \beta_2 q_2^2 = 0$.

Proof. Trivial. (Just apply Lemma 2.4.1 twice.)

This lemma is constructive in that it again yields a system of linear equations and a direct way of computing a C^1-interpolating quadric surface. Furthermore, a solution to the above equations, linear in the αs and βs, exists if and only if such an interpolating quadric surface exists. Again, when real surfaces are favorable, one requires α_1, α_2, β_1, and β_2 to be real numbers.

Example 2.4.1. Suppose $C_1 : (x^2 + z^2 - 1 = 0, 3x + y = 0)$, and $C_2 : (y^2 + z^2 - 1 = 0, x + 3y = 0)$. *The following equation is obtained from Lemma 2.4.2:* $(\alpha_1 + 9\beta_1 - \beta_2)x^2 + (\beta_1 - \alpha_2 - 9\beta_2)y^2 + (\alpha_1 - \alpha_2)z^2 + (6\beta_1 - 6\beta_2)xy + (\alpha_1 - \alpha_2) = 0$. *This implies* $\alpha_1 = \alpha_2$, $\beta_1 = \beta_2$, $\alpha_1 = -8\beta_1$. *When* $\alpha_1 = -8$ *and* $\beta_1 = 1$, *the interpolating surface is* $x^2 + y^2 - 8z^2 + 6xy + 8 = 0$.

In Lemma 2.4.2 and the example, the two conics on the given quadric surfaces, S_1 and S_2, were fixed. If one has freedom to choose different intersecting planes Q_1 and Q_2 then one is able to find a family of quadric interpolating surfaces. In this case, the equations of planes Q_1 and Q_2 would have unknown coefficients and the use of Lemma 2.4.2 would result in a nonlinear system of equations, linear in terms of α_1, α_2, β_1, and β_2, and quadratic in terms of the unknowns of the plane's equations.

Now, rather than trying to find a single quadric surface, one can also extend the above Lemma 2.4.2, to construct two or more quadrics that smoothly contain two given conics in space, and furthermore themselves intersect in a smooth fashion. The following Lemma 2.4.3, which is constructive, tells us how to go about this.

LEMMA 2.4.3. *Let* $C_1 : (f_1 = 0, q_1 = 0)$ *and* $C_2 : (f_2 = 0, q_2 = 0)$ *be two conics in space. These two curves can be smoothly contained by two "smoothly intersecting" quadrics* $S_1 : g_1 = a_1 f_1 + b_1 q_1^2 = 0$ *and* $S_2 : g_2 = a_2 f_2 + b_2 q_2^2$ *if and only if there exist nonzero constants* a_1, a_2, b_1, b_2, α, β, *and a plane* $Q : q(x, y, z) = 0$ *such that* $a_1 f_1 + b_1 q_1^2 - \alpha(a_2 f_2 + b_2 q_2^2) - \beta q^2 = 0$.

Proof. From Theorem 2.4.3 we note that two quadrics that intersect smoothly (at least C^1) must intersect with multiplicity at least two. It follows then from Bezout's Theorem 2.4.1 for surface intersection that the two quadrics S_1 and S_2 must meet in a plane curve (either an irreducible conic or straight lines). Let the intersection curve lie on the unknown plane Q, then just apply Lemma 2.4.1 three times.

The final equation of the above lemma results in a nonlinear (cubic) system of equations that is linear in terms of the unknowns a_1, a_2, b_1, b_2, α, and β, and quadratic in terms of the unknown coefficients of the plane $Q : q = 0$. Note that in Lemma 2.4.3 the quadric surfaces S_1 and S_2 need not be in the form given (as constructed via Lemma 2.4.1), but may instead be an m-parameter family of solutions, obtained by C^1-interpolation of input curves with possibly "normal" data, as explained in the prior sections.

The above method of Lemma 2.4.3 can straightforwardly be extended to finding a C^1-continuous mesh of k quadric surfaces that smoothly contain k conics in space.

THEOREM 2.4.4. *Let* $C_1 : (f_1 = 0, q_1 = 0)$, $C_2 : (f_2 = 0, q_2 = 0)$ $\cdots$ $C_k : (f_k = 0, q_k = 0)$ *be* k *conics in space. These curves can be smoothly contained by* k *quadrics* $S_1 : g_1 = a_1 f_1 + b_1 q_1^2 = 0$, $S_2 : g_2 = a_2 f_2 + b_2 q_2^2$, $\cdots$, $S_k : g_k = a_k f_k + b_k q_2^k$, *which themselves "smoothly intersect" if and only if there exist nonzero constants* a_1, a_2, $\cdots$, a_k, b_1, b_2, $\cdots$, b_k, α_1, $\cdots$, α_{k-1}, β_1, $\cdots$, β_{k-1} *and planes* $R_1 : r_1(x, y, z) = 0$, $\cdots$, $R_{k-1} : r_{k-1}(x, y, z) = 0$ *such that*

$$a_1 f_1 + b_1 q_1^2 - \alpha_1(a_2 f_2 + b_2 q_2^2) - \beta_1 r_1^2 = 0,$$
$$a_2 f_2 + b_2 q_2^2 - \alpha_2(a_3 f_3 + b_3 q_3^2) - \beta_2 r_2^2 = 0,$$
$$\cdots$$

$$a_{k-1} f_{k-1} + b_{k-1} q_{k-1}^2 - \alpha_{k-1}(a_k f_k + b_k q_k^2) - \beta_{k-1} r_{k-1}^2 = 0.$$

Proof. Direct applications of Lemma 2.4.3.

Note again as before, that in the above theorem, the quadric surfaces S_1, $\cdots$, S_k need not be in the form given (as constructed via Lemma 2.4.1), but may instead be an m-parameter family of solutions, obtained by C^1-interpolation of input curves with possibly "normal" data, as explained in the previous section. Also note, that given k conics in space, in general k quadrics above, may not form a C^1-continuous mesh (no nontrivial solution for the generated system of polynomial equations). In this case one may try increasing the number of quadric surface patches between any two of the given curves. This yields the theorem below, a variation of Theorem 2.4.4.

THEOREM 2.4.5. *Let* $C_1 : (f_1 = 0, q_1 = 0)$ *and* $C_2 : (f_2 = 0, q_2 = 0)$ *be two conics in space. These curves can be smoothly contained by two quadrics* $S_1 : g_1 = a_1 f_1 + b_1 q_1^2 = 0$, $S_2 : g_2 = a_2 f_2 + b_2 q_2^2$, *which together with* k *other quadrics* $T_1 : h_1 = 0$, $\cdots$, $T_k : h_k = 0$ *form a* C^1-*continuous mesh if and only if there exist nonzero constants* a_1, a_2, b_1, b_2, c_{i0}, $\cdots$, c_{i9} *(the coefficients of the quadric* $T_i : h_i = 0$*),* $i = 1 \cdots k$, *and* α_1, $\cdots$, α_{k+1}, β_1, $\cdots$, β_{k+1}, *and planes* $R_1 : r_1(x, y, z) = 0$, $\cdots$, $R_{k+1} : r_{k+1}(x, y, z) = 0$ *such that*

$$a_1 f_1 + b_1 q_1^2 - \alpha_1 h_1 - \beta_1 r_1^2 = 0,$$
$$a_2 f_2 + b_2 q_2^2 - \alpha_{k+1} h_k - \beta_{k+1} r_{k+1}^2 = 0,$$
$$h_i = \alpha_i h_{i-1} + \beta_i r_i^2, \quad i = 2, \cdots, k.$$

Necessarily the complexity of the nonlinear system of equations also goes up.

If the generated systems of polynomial equations in the above theorems do not yield a satisfactory C^1 solution, one may instead try intermixing cubic surfaces with quadrics. To do this one first considers the lemma below, similar to Lemma 2.4.1 and a corollary of Theorem 2.4.2.

LEMMA 2.4.4. *Let $S : f(x,y,z) = 0$ be an irreducible quadric surface, and $Q : q(x,y,z) = 0$ be a plane that intersects S in a conic C. Then, a cubic surface $T_1 : f_1(x,y,z)$ is tangent to S along C if and only if there exists nonzero constants $a_1, \cdots, a_4$, and $b_1, \cdots, b_4$ such that $f_1 = (a_1 x + a_2 y + a_3 z + a_4)f + (b_1 x + b_2 y + b_3 z + b_4)q^2$.*

Similar to Lemma 2.4.3 one obtains the following lemma.

LEMMA 2.4.5. *Let $C_1 : (f_1 = 0, q_1 = 0)$ and $C_2 : (f_2 = 0, q_2 = 0)$ be two conics in space. These two curves can be smoothly contained by two quadrics $S_1 : g_1 = a_1 f_1 + b_1 q_1^2 = 0$ and $S_2 : g_2 = a_2 f_2 + b_2 q_2^2 = 0$ both of which meet a cubic surface $T_1 : h_1 = 0$ if there exist nonzero constants $a_1,\ a_2,\ b_1,\ b_2,\ \alpha_{11}, \cdots, \alpha_{14},\ \alpha_{21}, \cdots, \alpha_{24}\ \beta_{11}, \cdots, \beta_{14},\ \beta_{21}, \cdots, \beta_{24}$ and planes $R_1 : r_1(x,y,z) = 0,\ R_2 : r_2(x,y,z) = 0$ such that $h_1 = (\alpha_{11} x + \alpha_{12} y + \alpha_{13} z + \alpha_{14})g_1 + (\beta_{11} x + \beta_{12} y + \beta_{13} z + \beta_{14})r_1^2 = (\alpha_{21} x + \alpha_{22} y + \alpha_{23} z + \alpha_{24})g_2 - (\beta_{21} x + \beta_{22} y + \beta_{23} z + \beta_{24})r_2^2$.*

Proof. It follows from Bezout's Theorem 2.4.1 for surface intersection that the a quadrics S_1 and a cubic surface T_1 must meet in either a space cubic, a plane cubic, an irreducible conic, or straight lines. Consider only the plane intersection curves and assume they lie on an unknown plane Q, then just apply Lemma 2.4.4.

In both the above lemmas, T_1 need not be in the above form but may instead be a l-parameter family of solutions, obtained by C^1-interpolation of input curves with possibly "normal" data, as explained in the previous section. These parameterized cubic surfaces may be intermixed with the quadric surfaces in Theorems 2.4.4 and 2.4.5 to form a C^1-continuous mesh of alternating quadric and cubic surfaces in the obvious manner.

2.4.4. Examples.

Example 2.4.2. A C^1 Mesh of a Family of Quadric and Quartic Surfaces. Consider a wireframe of a solid model consisting of two circles, $C_1 : (x^2 + y^2 + z^2 - 25 = 0, x = 0)$, and $C_2 : (x^2 + y^2 + z^2 - 25 = 0, y = 0)$. Each curve is associated with a "normal" direction that is chosen in the same direction as the gradients of the sphere. That is, $\mathbf{n}_1(x,y,z) = (0, 2y, 2z)$, and $\mathbf{n}_2(x,y,z) = (2x, 0, 2z)$. The wireframe has four faces to be fleshed, $\mathrm{face}_1 = (x \geq 0, y \geq 0)$, $\mathrm{face}_2 = (x \geq 0, y \leq 0)$, $\mathrm{face}_3 = (x \leq 0, y \leq 0)$, and $\mathrm{face}_4 = (x \leq 0, y \geq 0)$.

In Fig. 2.8, face_1 and face_3 are filled with the patches taken from the sphere $x^2 + y^2 + z^2 - 25 = 0$. To flesh the remaining faces with overall C^1-continuity along all interpatch boundary curves requires degree 4 surface patches. Using the interpolation algorithms of §2.2 yields C_1 and C_2, both 11-parameter (homogeneous) family of quartic $\mathbf{G}^1$ interpolating surfaces, given by $f(x,y,z) = r_1 z^4 + (r_2 y + r_6 x + 5r_4)z^3 + (r_3 y^2 + (r_7 x + 5r_8)y + r_{10} x^2 + 5r_{11} x - 25r_9 - 25r_1)z^2 + (r_2 y^3 + (r_6 x + 5r_4)y^2 + (r_2 x^2 - 25r_2)y + r_6 x^3 +$

$5r_4x^2 - 25r_6x - 125r_4)z + (r_3 - r_1)y^4 + (r_7x + 5r_8)y^3 + (r_5x^2 + 5r_{11}x - 25r_9 - 25r_3 + 25r_1)y^2 + (r_7x^3 + 5r_8x^2 - 25r_7x - 125r_8)y + (r_{10} - r_1)x^4 + 5r_{11}x^3 + (-25r_9 - 25r_{10} + 25r_1)x^2 - 125r_{11}x + 625r_9$. An instance from this family is $f(x,y,z) = -1250 - x^4 - y^4 - x^2z^2 - y^2z^2 + 50z^2 + 75y^2 + 75x^2$ used to fill faces $face_2$ and $face_4$ in Fig. 2.8.

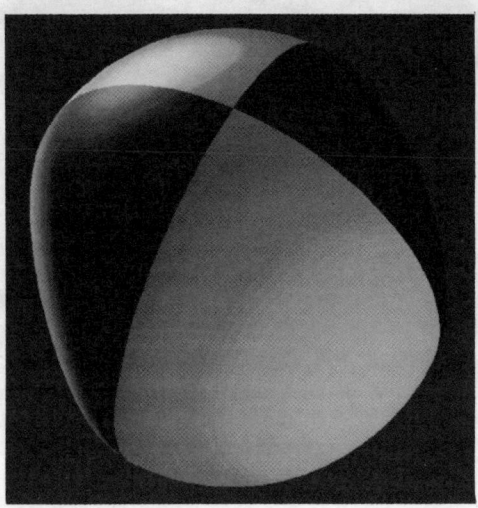

FIG. 2.8. C^1 *mesh of quadric and quartic patches.*

Example 2.4.3. A C^1 Mesh of Quadric Patches. Let conic C_1 be given by $f_1 = x^2 + y^2 - z^2 + 4xy + 4x + 4y + 3 = 0$ (a hyperboloid of one sheet) and $q_1 = x + y + 1 = 0$. Similarly, let conic C_2 be given by $f_2 = 19x^2 + 10y^2 - 9z^2 + 38xy - 114x - 114y + 180 = 0$ (a hyperboloid of one sheet), $q_2 = x + y - 3 = 0$, and let the unknown plane be $P : ax + by + cz + d = 0$. Then the equation for the system of smooth interpolating quadrics $a_1f_1 + b_1q_1^2 - \alpha(a_2f_2 + b_2q_2^2) = \beta(ax + by + cz + d)^2$ results in a nonlinear system of 10 equations: $-\beta c^2 + 9a_2\alpha - a_1 = 0$, $-2b\beta c = 0$, $-2a\beta c = 0$, $-2\beta cd = 0$, $-b^2\beta - \alpha b_2 + b_1 - 10a_2\alpha + a_1 = 0$, $-2ab\beta - 2\alpha b_2 + 2b_1 - 38a_2\alpha + 4a_1 = 0$, $-2b\beta d + 6\alpha b_2 + 2b_1 + 114a_2\alpha + 4a_1 = 0$, $-a^2\beta - \alpha b_2 + b_1 - 19a_2\alpha + a_1 = 0$, $-2a\beta d + 6\alpha b_2 + 2b_1 + 114a_2\alpha + 4a_1 = 0$, and $-\beta d^2 - 9\alpha b_2 + b_1 - 180a_2\alpha + 3a_1 = 0$. This nonlinear system has a nontrivial solution (in the sense that a_1, a_2, and α are nonzero) : $a_1 = -a^2\beta$, $b_1 = 2a^2\beta$, $a_2 = -\frac{a^2\beta}{9\alpha}$, $b_2 = \frac{19a^2\beta}{9\alpha}$, and $b = c = d = 0$.[5] Hence, the two conics C_1 and C_2 are smoothly contained by quadrics $g_1 = 0$ and $g_2 = 0$, respectively, and which in turn, smoothly intersect in a conic in the plane Q. The real quadric $g_1 = x^2 + y^2 + z^2 - 1 = 0$ is a sphere, while the other real quadric $g_2 = y^2 + z^2 - 1$ is a cylinder. Note that the above solution implies that there is only one pair of real quadric surfaces that smoothly contain the given conics. Also, for this case it can be shown that neither a single quadric nor a single cubic surface can Hermite interpolate

---

[5]This nonlinear system was solved with the aid of MACSYMA, on a Symbolics 3650.

the two given conics. Geometrically then, the two hyperboloids of one sheet are smoothly joined by a sphere and a cylinder. See Fig. 2.9.

FIG. 2.9. C^1 *mesh of quadric patches.*

2.4.5. Open Problem.

1. Extend the technique of constructing C^1-continuous meshes to constructing C^k-continuous meshes using the definition of C^k-continuity of §2.2 and the theorems of the above section.

2.5. Triangulated Data Fit with Surface Patches

2.5.1. Problem.   Given a triangulation $\mathcal{T}$ of points $\mathbf{p} = (x_i, y_i, z_i)$ in $\mathbb{R}^3$, possibly with various order derivatives at these points, construct a C^k-continuous mesh of real algebraic surface patches S_i, respecting the topology of the triangulation $\mathcal{T}$, and which C^k-interpolates the collection of points $\mathbf{p}_j = (x_j, y_j, z_j)$ in $\mathbb{R}^3$ and derivatives at those points. Additionally, this C^k mesh of patches may also C^l least-squares approximate a collection of points $\mathbf{q}_k = (x_k, y_k, z_k)$ and derivatives in $\mathbb{R}^3$.

2.5.2. Summary of Approaches.   The generation of a mesh of smooth surface patches or *splines* that interpolate or approximate *triangulated space data* is one of the central topics of geometric design. Chui [20] and De Boor [22] summarize much of the history of previous work. Interpolatory spline problems can be classified by the following factors:

- What kind of surface patches are used for a basis: parametric or implicit?

- What kind of data triangulation exists: functional values over two-dimensional triangulation or arbitrarily spaced three-dimensional triangulation?

- What kinds of information is provided as input data : C^0 data, C^1 data, C^2 data, and so on?

- With what order of geometric continuity are the patches required to meet along the boundary curves?

- Is the interpolant required to be local, that is, each patch is to be constructed only from nearby data or global information?

- Do they split one macro triangle into many micro-triangles or not?

- How ill-conditioned is the solution to perturbations in the input data?

- How efficient are the algorithms?

Prior work on splines have traditionally worked with a given planar triangulation using a polynomial function basis or, alternatively, a triangulation in three dimensions using a parametric surface for each triangular face [3], [13], [15], [16], [19], [24], [31], [32], [34], [36], [43], [44], [48], [56], [57], [59]. An exception is [8] which computes the dimension of implicit splines over a planar or spatial triangulation for an implicit algebraic surface replacing each triangular face.

Little work has been done on the construction of explicit spline basis for implictly defined algebraic surfaces. Sederberg [55] shows how various smooth implicit surfaces can be manipulated as functions in Bézier control tetrahedra with finite weights. Dahmen [21] presents the construction of tangent plane continuous piecewise triangular quadric surfaces. In his construction a macro patch is split into six micro-quadratic patches, similar to the splitting scheme of Powell and Sabin [49]. The resulting surfaces interpolate finite sets of essentially arbitrary points in $\mathbf{R}^3$ according to a given topology and given normal directions at the points within some ranges depending on the topology and the location of data points. The technique, however, works only if the original triangulation of the data set allows a transversal system of planes and hence is quite restricted. Moore and Warren [42] extend the planar C^0-approximation scheme of [39] and compute a C^1 piecewise quadratic approximation to scattered data. Bajaj and Ihm [10] use a single implicit surface for each macro patch at the expense of a higher degree 5 surface. This quintic surface provides sufficient flexibility in globally C^1 surface fitting as well as provides local shape control. These surface patches are used in the reconstruction of smooth models of the human anatomy from CT/MRI data [7].

2.5.3. Recent Results. The method of [10] takes as input a three-dimensional triangulation $\mathcal{T}$ of points v_i in $\mathbf{R}^3$ with possibly first-order derivative information ("normals") at the points. From the input, a wireframe mesh of conic curves is first constructed. Each conic replaces an edge e_j of the triangulation $\mathcal{T}$ and C^1 interpolates the end points and specified normals at those points. If no normals are specified, a unique normal is chosen, for example, by averaging the normals of the incident faces f_k of T at that point. Unique normals, or in other words, unique tangent planes at the end points is a necessary conditon for global C^1 fitting. Each conic, being a rational curve, is represented by its rational parameterization $E_j(t) = (\frac{x_j(t)}{w_j(t)}, \frac{y_j(t)}{w_j(t)}, \frac{z_j(t)}{w_j(t)})$. Next, along each such conic edge a quadratically varying "normal" $N_j(t) = (\frac{nx(t)}{w(t)}, \frac{ny(t)}{w(t)}, \frac{nz(t)}{w(t)})$ is specified, with the property that $E_j'(t) \cdot N_j(t)$ is identically zero. The $N_j(t)$ necessarily take on the same values as the unique normals at the end points.

Lee [38] presents a compact method for computing the above rational quadratic parametric representation for each of the edge conics. In particular, a ρ-conic parameterization is derived in which the remaining one degree of freedom of each conic edge is controlled in terms of the ρ value. The resulting topology of the wireframe is that of the initial triangulation T, with each triangular face F_k now consisting of conic edges. The shape of each conic edge in turn being controlled by an independent ρ parameter.

Bajaj and Ihm [10] provide a local C^1 implicit surface interpolant S_k of degree 5 which fills each curvilinear triangular facet F_k. The degree 5 implicit surface suffices because its degrees of freedom are 55 while the maximum number of C^1-interpolatory constraints of F_k is 51. Each interpolant S_k has thus four independent parameters of local freedom, which are used for shape control. Further details on interactive shape control are in §2.6.

2.5.4. Examples.

Example 2.5.1. *Locally Supported Triangular C^1-Interpolants for Smoothing Polyhedra.* The input is an octahedron represented by vertices $v_i, i = 0 \cdots 5$, edges $e_j, j = 0 \cdots 11$ and faces $f_k, k = 0 \cdots 7$, together with vertex normals $n_l, l = 0 \cdots 5$.

$$v_0 = (0.0, 0.0, 0.0), \qquad v_1 = (2.0, 4.0, 0.0),$$
$$v_2 = (-0.1, 4.0, 2.1), \qquad v_3 = (2.0, 3.0, 2.0),$$
$$v_4 = (1.0, 0.0, 2.0), \qquad v_5 = (2.5, 0.0, 1.0).$$

$$e_0 = (v_4, v_0), \qquad e_1 = (v_5, v_4),$$
$$e_2 = (v_0, v_5), \qquad e_3 = (v_5, v_1),$$
$$e_4 = (v_3, v_5), \qquad e_5 = (v_1, v_3),$$
$$e_6 = (v_3, v_4), \qquad e_7 = (v_2, v_4),$$
$$e_8 = (v_3, v_2), \qquad e_9 = (v_0, v_1),$$
$$e_{10} = (v_1, v_2), \qquad e_{11} = (v_2, v_0).$$

$$f_0 = (e_0, e_4, e_5), \qquad f_1 = (e_1, e_5, e_3),$$
$$f_2 = (e_5, e_4, e_3), \qquad f_3 = (e_4, e_2, e_3),$$
$$f_4 = (e_1, e_0, e_5), \qquad f_5 = (e_1, e_3, e_2),$$
$$f_6 = (e_2, e_4, e_0), \qquad f_7 = (e_2, e_0, e_1).$$

$$n_0 = (-0.592524, -0.557271, -0.581691),$$
$$n_1 = (0.573733, 0.581132, -0.577162),$$
$$n_2 = (-0.593023, 0.485480, 0.642365),$$
$$n_3 = (0.633794, 0.243658, 0.734122),$$
$$n_4 = (-0.104040, -0.537266, 0.836971),$$
$$n_5 = (0.840500, -0.541705, -0.010696).$$

First a wireframe of conics is constructed where each conic replaces an edge and C^1-interpolates the corresponding vertices of the edge. Next normals are constructed for each curvilinear conic edge of the wireframe and varying quadratically along the conics. Since the normals are quadratic functions and take on the value of the given normals at the vertex corners, specifying an additional normal vector at an interior point of each edge suffices.

$$\text{edgenorm}_0 = (-0.706837, 0.612762, 0.353418),$$
$$\text{edgenorm}_1 = (0.426401, 0.639602, 0.639602),$$
$$\text{edgenorm}_2 = (0.285582, 0.639307, -0.713954),$$
$$\text{edgenorm}_3 = (-0.515202, -0.267929, -0.814114),$$
$$\text{edgenorm}_4 = (-0.477035, -0.348458, 0.806855),$$
$$\text{edgenorm}_5 = (-0.608621, 0.709693, 0.354847),$$
$$\text{edgenorm}_6 = (-0.898131, 0.299377, 0.322078),$$
$$\text{edgenorm}_7 = (-0.828637, -0.214953, -0.516872),$$
$$\text{edgenorm}_8 = (0.339020, 0.766491, -0.545488),$$
$$\text{edgenorm}_9 = (-0.874728, 0.437364, 0.208721),$$
$$\text{edgenorm}_{10} = (-0.673887, -0.302905, -0.673887),$$
$$\text{edgenorm}_{11} = (0.158349, 0.462070, -0.872592).$$

See Fig. 2.10 (left) where the conics are constructed for the ρ value of 0.7.

The interpolation algorithm of the above section then constructs triangular C^1-interpolants—a four-parameter family of quintic surfaces, one family per triangular facet of the wireframe. Instances of quintic surface patches generated for this example are displayed in Fig. 2.10 (right).

2.5.5. Open Problems.

1. Build a quadratic wireframe for arbitrarily shaped three-dimensional triangulated data.

2. Provide interactive shape control for a p-parameter family of implicit surfaces.

3. Provide trade-off bounds of degree of implicit algebraic surfaces and the number of split patches in C^k interpolation.

2.6. Interactive Shape Control of Implicit Surface Patches

2.6.1. Problem. Interactively control the shape of an interpolating or approximating implicit surface by selecting appropriate instances from a p-parameter family of solution surfaces.

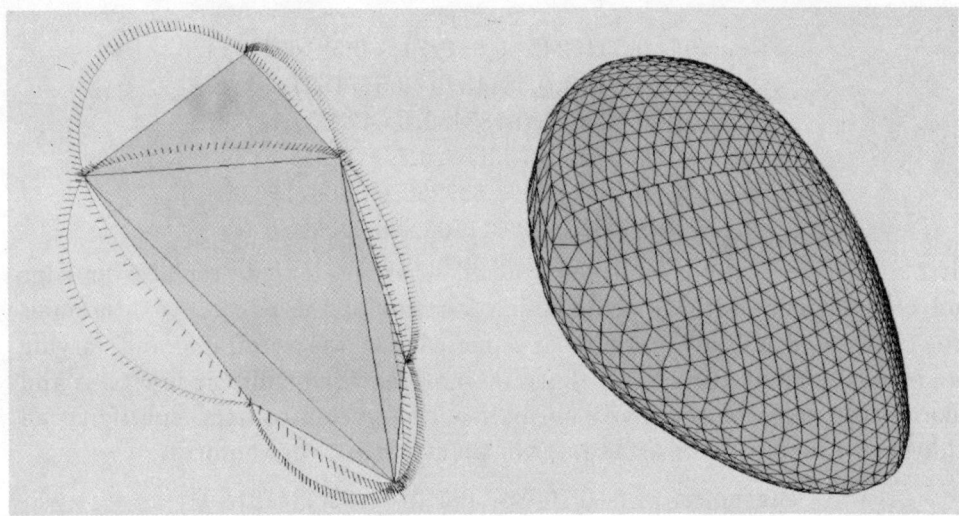

FIG. 2.10. *Locally supported triangular C^1-interpolants for smoothing a polyhedron.*

2.6.2. Summary of Approaches. The problem of interactively selecting a surface instance from a p-parameter family of solutions is equivalent to assigning values to each of the p parameters. When there are p parameters to be instantiated, one may additionally specify a set of points, curves, or even surfaces around the earlier given input data, which approximately describes the final surface to be designed. The final solution instance is computed via interpolation of the given input data and with least-squares approximation of the additional data set. This scheme is presented in the paper by Bajaj, Ihm, and Warren [12]. An example of this use is presented in Example 2.4.1 of §2.4.

An alternate scheme based on Sederberg's Bézier formulation of algebraic surfaces [55] is used for shape control of a family of surfaces in Bajaj and Ihm [11]. They present a method that allows a surface designer to intuitively and interactively control the shape of a C^k-interpolating or approximating surface, thereby choosing an appropriate instance from the family by automatically selecting values for the p distinct parameters.

2.6.3. Recent Results. The result of multivariate C^k-interpolation algorithm for a given data set is in general a p-parameter family of algebraic surfaces $f(x, y, z) = 0$, satisfying the given geometric data constraints. A surface designer must be able to choose an appropriate instance from this family, to satisfy his or her application by specifying values for the p parameters (say $\mathbf{r} = r_1, r_2, \cdots, r_p$). The equation for the family has the form

$$(2.2) \qquad f(x, y, z) = \sum_{i=0}^{n} \sum_{j=0}^{n-i} \sum_{k=0}^{n-i-j} c_{ijk} \cdot x^i y^j z^k = 0,$$

where c_{ijk} is a homogeneous linear combination of $\mathbf{r}$. Various distinct choices of values for $\mathbf{r}$ yield interpolating surface instances possessing different shapes. Bajaj and Ihm [11] present a method that allows a surface designer to intuitively and interactively control the shape of a Hermite interpolating surface, thereby choosing an appropriate instance from the family by automatically selecting values for the p distinct parameters.

The essential idea is to consider the interpolating family f as the zero contour $w = 0$ of the trivariate function $w = f(x, y, z)$. See Sederberg [55] where the same idea is used to define algebraic surface patches. Of course, since one considers a family of interpolating algebraic surfaces, the coefficients of f here have indeterminates r_i. The trivariate function, when transformed into barycentric coordinates, yields a control polyhedron with weights (the interactive control given to the designer). For the purpose here, the trivariate polynomial $f(x, y, z)$ is symbolically converted into a polynomial $F(s, t, u)$ in barycentric coordinates, specified over a tetrahedron. To concentrate on a specific portion of the algebraic surface, the designer appropriately chooses the location of the four vertices of the tetrahedron, enclosing the desired region. Shape of the Hermite interpolating surface is now controlled by changing the weights of control points associated with the tetrahedron.

Let the trivariate barycentric coordinates of points inside a tetrahedron be given by (s, t, u). The tetrahedron is specified by the designer who selects the location of its four vertices P_{n000}, P_{0n00}, P_{00n0}, and P_{000n}. The Cartesian coordinates P of a point inside the tetrahedron are related to its barycentric coordinates s, t, u by $P = sP_{n000} + tP_{0n00} + uP_{00n0} + (1 - s - t - u)P_{000n}$. Control points on the tetrahedron are defined by $P_{ijk} = \frac{i}{n}P_{n000} + \frac{j}{n}P_{0n00} + \frac{k}{n}P_{00n0} + \frac{n-i-j-k}{n}P_{000n}$ for nonnegative integers i, j, k such that $i + j + k \leq n$. With each control point there is also associated a weight w_{ijk}, corresponding to the coefficients c_{ijk} of (2.2), which is a linear (not necessarily homogeneous) combination of $\mathbf{r}$. All this together defines the p-parameter algebraic surface family in barycentric coordinates,

$$F(s, t, u) = \sum_{i=0}^{n}\sum_{j=0}^{n-i}\sum_{k=0}^{n-i-j} w_{ijk} \cdot \binom{n}{i, j, k} \cdot s^i t^j u^k (1 - s - t - u)^{n-i-j-k} = 0.$$

There are $\binom{n+3}{3}$ w_{ijk}, exactly as many as the c_{ijk}. Straightforward methods exist to converting a trivariate polynomial in the power basis with cartesian coordinates, to the form above in trivariate barycentric coordinates defined over a given tetrahedron.

Consider, as a simple example, a quadric surface which Hermite interpolates a line $LN : (1 - t, t, 0)$ with a normal $(0, 0, 1)$. Hermite interpolation algorithm returns a five-parameter family of surfaces $f(x, y, z)$ as in (2.2) with $n = 2$ and where $c_{200} = r_1$, $c_{110} = 2r_1$, $c_{101} = r_4$, $c_{100} = -2r_1$, $c_{020} = r_1$, $c_{011} = r_5$, $c_{010} = -2r_1$, $c_{002} = r_3$, $c_{001} = r2$, and $c_{000} = r_1$. For a given tetrahedron with vertices $P_{n00} = (2, 0, 0)$, $P_{0n0} = (0, 2, 0)$, $P_{00n} = (0, 0, 2)$, and $P_{000} = (0, 0, 0)$, the surface family representation $f(x, y, z)$ is transformed to $F(s, t, u)$ as in the equation above with $n = 2$ and where $w_{000} = r_1$, $w_{001} = r_1 + r_2$, $w_{002} = r_1 + 2r_2 + 4r_3$, $w_{010} = -r_1$, $w_{011} = -r_1 + r_2 + 2r_5$, $w_{020} = r_1$, $w_{100} = -r_1$, $w_{101} = -r_1 + r_2 + 2r_4$, $w_{110} = r_1$, and $w_{200} = r_1$

As $F(s, t, u) = 0$ in barycentric coordinates describes a constrained p-parameter family of algebraic surface patches of degree n, the change of one of the weights w_{ijk} associated with a control point of the tetrahedron affects the weights of other control points. For example, suppose $w_1 = r_1 + r_2 + r_3 + 2r_4 - 1$, $w_2 = r_1 + r_2 + r_4 + 5$, and $w_3 = r_3 + r_4$. Then we can derive the following linear relation between the weights: $w_1 - w_2 - w3 - 6 = 0$. (For notational simplicity we here consider, wlg, the weights w_{ijk} to be also indexed by a single parameter l, i.e., weights w_l). From this invariant, we get $\Delta w_1 - \Delta w_2 - \Delta w_3 = 0$, and every time weights are changed, the above invariant is maintained.

Hence, in general one derives a system of invariant equations

$$I_1(\Delta w_1, \Delta w_2, \cdots, \Delta w_c) = 0,$$
$$I_2(\Delta w_1, \Delta w_2, \cdots, \Delta w_c) = 0,$$
$$\vdots$$
$$I_l(\Delta w_1, \Delta w_2, \cdots, \Delta w_c) = 0,$$

from the linear weight expressions

$$w_1(r_1, r_2, \cdots, r_p) = w_1,$$
$$w_2(r_1, r_2, \cdots, r_p) = w_2,$$
$$\vdots$$
$$w_c(r_1, r_2, \cdots, r_p) = w_c.$$

This is easily achieved by Gaussian elimination. Changing some weights can now be considered as moving from a weight vector $\mathbf{W} = (w_1, w_2, \cdots, w_c)$ to another $\mathbf{W}' = (w_1', w_2', \cdots, w_c')$, with the constraint that $\Delta \mathbf{W} = \mathbf{W}' - \mathbf{W}$ is a solution of the computed system of invariant equations. Next, suppose that a surface designer wants to see how the surface shape changes with a value change of w_1 alone. However, a change in value of w_1 automatically changes the value of additional w_is related to it by the invariants I_js. Usually, the linear system of invariant equations are underdetermined, yielding an infinite number of choices of Δw_is $(i = 2, 3, \cdots, c)$. Then how does the designer make a choice of values for the w_is that reflects the influence of a change of only w_1, as clearly as possible?

One possible heuristic is to minimize the two-norm of $(\Delta w_2, \cdots, \Delta w_c)$, and hence the two-norm of $\Delta \mathbf{W}$. Note that $\|\Delta \mathbf{W}\|_2^2 = \Delta w_1^2 + \Delta w_2^2 + \cdots + \Delta w_c^2$. For a change $\Delta w_1 = d$, one sees that

$$I_1(d, \Delta w_2, \cdots, \Delta w_c) = 0,$$
$$I_2(d, \Delta w_2, \cdots, \Delta w_c) = 0,$$

$$\vdots$$

$$I_l(d, \Delta w_2, \cdots, \Delta w_c) = 0,$$

will have a solution $\Delta \mathbf{W}^0 = (d, \Delta w_2^0, \cdots, \Delta w_c^0)$ where Δw_i^0s are expressed linearly through another set of free parameters $q_1, q_2, \cdots, q_{p-1}$. Hence, $\|\Delta \mathbf{W}^0\|_2$ is a quadratic function of the new parameters, which we denote by $Q(q_1, q_2, \cdots, q_{p-1})$.

In order to minimize the norm of $\Delta \mathbf{W_0}$, the quadratic function $Q(q_1, q_2, \cdots, q_{p-1})$ needs to be minimized. Since Q is quadratic, the minimum solution can be obtained straightforwardly by solving the linear system $\nabla Q(q_1, q_2, \cdots, q_{p-1}) = \mathbf{0}$. If the (unique) minimum solution point is $Q^0 = (q_1^0, q_2^0, \cdots, q_{(p-1)}^0)$, then $\Delta \mathbf{W}^0 = (d, \Delta w_2^0, \cdots, \Delta w_c^0)$ corresponding to Q^0 will define the desired change of weights of $w_2, \cdots, w_n$ having minimum effect on the shape of the surface. The instance surface for the new weights $\mathbf{W}' = \mathbf{W} + \Delta \mathbf{W}^0$ will then reflect predominantly the effect of the change of w_1 by $\Delta w_1 = d$.

2.6.4. Examples.

Example 2.6.1. *Interactive Shape Control of a Family of Quartic Surfaces.* Construct the lowest degree surface that can smoothly join three truncated orthogonal circular cylinders $CYL_1 : x^2 + y^2 - 1 = 0$ for $z \geq 2$, $CYL_2 : y^2 + z^2 - 1 = 0$ for $x \geq 2$, and $CYL_3 : z^2 + x^2 - 1 = 0$ for $y \geq 2$. C^1-interpolation shows that the minimum degree for such a joining surface is four, and the null space of the interpolation matrix is a two homogenous parameter (or one independent affine parameter) family of algebraic surfaces. Consider a circle $C_1 : (\frac{2t}{1+t^2}, \frac{1-t^2}{1+t^2}, 2)$ on CYL_1 with the associated rational "normal" $\mathbf{n}_1(t) :$ $(\frac{4t}{1+t^2}, \frac{2-2t^2}{1+t^2}, 0)$, the circle $C_2 : (2, \frac{2t}{1+t^2}, \frac{1-t^2}{1+t^2})$ on CYL_2 with the associated rational "normal" $\mathbf{n}_2(t) : (0, \frac{4t}{1+t^2}, \frac{2-2t^2}{1+t^2})$, and the circle $C_3 : (\frac{2t}{1+t^2}, 2, \frac{1-t^2}{1+t^2})$ on CYL_3 with the associated rational "normal" $\mathbf{n}_3(t) : (\frac{4t}{1+t^2}, 2, \frac{2-2t^2}{1+t^2})$.

Again, all C_1, C_2, and C_3s "normals" are respectively chosen in the same direction as the gradients of their corresponding containing surfaces CYL_1, CYL_2, and CYL_3. This ensures that any interpolating surface for C_1, C_2, and C_3 will also meet CYL_1, CYL_2, and CYL_3 smoothly along these curves. A degree three algebraic surface does not suffice since the rank of the resulting linear system is greater than 19, the number of independent unknowns. Next as a possible interpolant one considers a degree four algebraic surface with 34 independent unknown coefficients. Repeating the interpolation for the original curves results in 52 equations. The rank of this linear system is 33, and thus there is a one affine independent parameter family of quartic Hermite interpolating surface, which is $f(x, y, z) = r_1 z^4 + \frac{r_2+10r_1}{12} yz^3 + \frac{r_2+10r_1}{12} xz^3 - \frac{r_2+10r_1}{3} z^3 + 2r_1 y^2 z^2 + \frac{r_2+10r_1}{12} xyz^2 - \frac{r_2+10r_1}{3} yz^2 + 2r_1 x^2 z^2 - \frac{r_2+10r_1}{3} xz^2 + r_2 z^2 + \frac{r_2+10r_1}{12} y^3 z + \frac{r_2+10r_1}{12} xy^2 z - \frac{r_2+10r_1}{3} y^2 z + \frac{r_2+10r_1}{12} x^2 yz - \frac{r_2+10r_1}{3} xyz + \frac{r_2+10r_1}{4} yz + \frac{r_2+10r_1}{12} x^3 z - \frac{r_2+10r_1}{3} x^2 z + \frac{r_2+10r_1}{4} xz + \frac{r_2+10r_1}{3} z + r_1 y^4 + \frac{r_2+10r_1}{12} xy^3 - \frac{r_2+10r_1}{3} y^3 +$

$2r_1x^2y^2 - \frac{r_2+10r_1}{3}xy^2 + r_2y^2 + \frac{r_2+10r_1}{12}x^3y - \frac{r_2+10r_1}{3}x^2y + \frac{r_2+10r_1}{4}xy + \frac{r_2+10r_1}{3}y +$
$r_1x^4 - \frac{r_2+10r_1}{3}x^3 + r_2x^2 + \frac{r_2+10r_1}{3}x + \frac{5r_1-7r_2}{3}.$

An instance of this family ($r_1 = 1$, $r_2 = 10$) is shown in Fig. 2.11. This figure illustrates three different instances of $f(x, y, z) = 0$ obtained by changing the value of w_{000}. Each time w_{000} is varied, the invariant equations are met, and each instance surface Hermite $\mathbf{G}^1$-interpolates the three input curves. As the value of w_{000} continually increases from $w_{000} < 0$, the surface eventually passes through $P_{000} = (0, 0, 0)$ for $w_{000} = 0$, and then separates into three irreducible parts for $w_{000} > 0$.

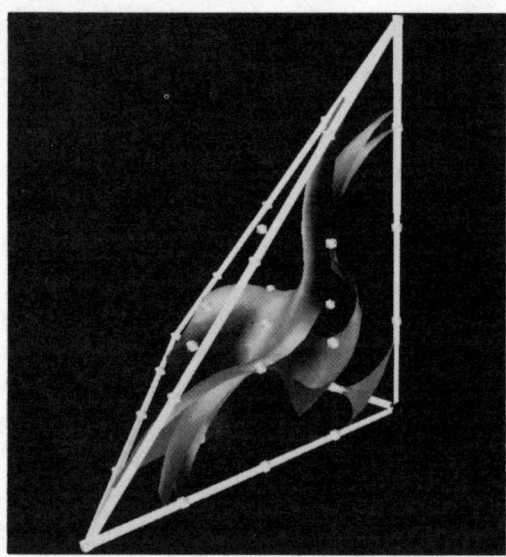

FIG. 2.11.   *Interactive shape control of a family of quartic surfaces.*

2.6.5. Open Problem.

1. Derive other intuitive techniques for shape control.

2.7. Conclusion

Many of the interpolation and approximation algorithms presented here were implemented by Insung Ihm as part of SHILP, our solid modeling and display system [4]. The programs takes as input any collection of geometric data points, curves, with and without associated "normals" and their derivatives. Both implicit and rational parametric representations of the space curves and their derivatives are allowed. The rank computation is done implicitly during the solution steps. The eigenvalue computation for least square computations is done with the help of routines from EISPACK. The result, when nontrivial solutions exist, are expressed in terms of symbolic coefficients and represent a family of interpolation surfaces. Values are specified for these coefficients by means of either the least-squares approximation approach as indicated in §2.3, or using Bézier control weights as detailed in §2.6. Desirable nonsingular and irreducible, real algebraic surfaces are computed. We are currently improving this implementation to include a more user-friendly method of instantiating the interpolated solution as well as a way of automatically incorporating the nonsingular and irreducibility constraints [35].

Acknowledgment

I am grateful to my graduate student Insung Ihm for his help with the pictures.

References

[1] A. Albano, *Representation of digitized contours in terms of conic arcs and straight line segments*, Computer Graphics and Image Processing, 3 (1974), pp. 23–33.

[2] ——, *Scattered data interpolation in three or more variables*, in Mathematical Methods in Computer Aided Geometric Design, T. Lyche and L. Schumaker, eds., Academic Press, New York, 1989, pp. 1–34.

[3] P. Alfeld, B. Piper, and L. Schumaker, *Minimally supported bases for spaces of bivariate piecewise polynomials of smoothness r for degree $d > 4r + 1$*, Comput. Aided Geom. Des., 4 (1987), pp. 105–123.

[4] V. Anupam, C. Bajaj, T. Dey, and I. Ihm, *The SHILP Solid Modeling and Display Toolkit*, Computer Science Tech. Report, CSD-TR-91-064, CAPO-91-29, Purdue University, 1991.

[5] C. Bajaj, *Geometric modeling with algebraic surfaces*, in The Mathematics of Surfaces III, D. Handscomb, ed., Oxford University Press, Oxford, 1989, pp. 3–48.

[6] ——, *C^1 interpolation using piecewise quadrics and cubic surfaces*, in Curves and Surfaces in Computer Graphics and Vision, Proc. of the Symposium on Electronic Imaging Science & Technology, Santa Clara, CA, 1990, pp. 82–93.

[7] ——, *Electronic skeletons: Modeling skeletal structures with piecewise algebraic surfaces*, in Curves and Surfaces in Computer Graphics and Vision II, Proc. of the Symposium on Electronic Imaging Science & Technology, 1610, Boston, MA, 1991, pp. 230–237.

[8] C. Bajaj and J. Hopcroft, *The combinatorics of algebraic splines*, manuscript.

[9] C. Bajaj and I. Ihm, *Hermite interpolation using real algebraic surfaces*, Proc. of the Fifth ACM Symposium on Computational Geometry, West Germany, 1989, pp. 94–103.

[10] ——, *C^1 smoothing of polyhedra using implicit surface patches*, Proc. of the ACM SIGGRAPH '92, Chicago, IL, 1992.

[11] ——, *Algebraic surface design with Hermite interpolation*, ACM Trans. Graphics, 19 (1992), pp. 61–91.

[12] C. Bajaj, I. Ihm, and J. Warren, *Exact and least-squares approximate G^k fitting of implicit algebraic surfaces*, ACM Trans. Graphics (1992), to appear.

[13] E. Beeker, *Smoothing of shapes designed with free-form surfaces*, Comput. Aided Des., 18 (1986), pp. 224–232.

[14] R. Biggerstaff, *Three variations in dental arch form estimated by a quadratic equation*, J. Dental Res., 51 (1972), p. 1509.

[15] L. Billera, *Homology of smooth splines: Generic triangulations and a conjecture of Strang*, Trans. Amer. Math. Soc., 310 (1988), pp. 325–340.

[16] L. Billera and L. Rose, *Grobner basis methods for multivariate splines*, Mathematical Methods in Computer Aided Geometric Design, T. Lyche and L. Schumaker, eds., Academic Press, New York, 1989, pp. 93–104.

[17] F. L. Bookstein, *Fitting conic sections to scattered data*, Computer Graphics and Image Processing, 9 (1979), pp. 56–71.

[18] W. Boehm, G. Farin and J. Kahmann, *A survey of curves and surface methods in CAGD*, Comput. Aided Geom. Des., 1 (1984), pp. 1–60.

[19] H. Chiyokura and F. Kimura, *Design of solids with free-form surfaces*, Computer Graphics, 17 (1983), pp. 289–298.

[20] C. Chui, *Multivariate Splines*, CBMS-NSF Regional Conference Series in Applied Mathematics, 54, Society for Industrial and Applied Mathematics, Philadelphia, 1988.

[21] W. Dahmen, *Smooth piecewise quadric surfaces*, in Mathematical Methods in Computer Aided Geometric Design, T. Lyche and L. Schumaker, eds., Academic Press, New York, 1989, pp. 181–194.

[22] C. De Boor, *Quasi-interpolants and approximation power of multivariate splines*, in Computation of Curves and Surfaces, W. Dahmen, M. Gasca, and C. Michelli, eds., Kluwer Academic Pub., 1990, pp. 313–346.

[23] T. DeRose, *Geometric continuity: A parameterization independent measure of continuity for computer aided design*, Ph.D. Thesis, Computer Science, University of California, Berkeley, 1985.

[24] G. Farin, *Triangular Berstein-Bézier patches*, Comput. Aided Geom. Design., 3 (1986), pp. 83–127.

[25] P. Fjallstrom, *Smoothing of polyhedral models*, in Proc. of the Second ACM Symposium on Computational Geometry, Yorktown Heights, NY, 1986, pp. 226–235.

[26] D. Franke, *Scattered data interpolation: Tests of some methods*, Math. Comp., 38 (1982), pp. 181–200.

[27] W. Fulton, *Intersection Theory*, Springer-Verlag, New York–Berlin, 1984.

[28] T. Garrity, and J. Warren, *Geometric continuity*, Computer Science Technical Report TR88-89, Rice University, 1989.

[29] R. Gnanadesikan, *Methods for Statistical Data Analysis of Multivariate Observations*, John Wiley, New York, 1977.

[30] G. Golub, and R. Underwood, *Stationary values of the ratio of quadratic forms subject to linear constraints*, Z. Agnew. Math. Phys., 21 (1970), pp. 318–326.

[31] J. Gregory and P. Charrot, *A C^1 triangular interpolation patch for computer aided geometric design*, Computer Graphics and Image Processing, 13 (1980), pp. 80–87.

[32] H. Hagen, and H. Pottmann, *Curvature continuous triangular interpolants*, in Mathematical Methods in Computer Aided Geometric Design, T. Lyche and L. Schumaker, eds., Academic Press, New York, 1989, pp. 373–384.

[33] G. Herron, *Smooth closed surfaces with discrete triangular interpolants*, Comput. Aided Geom. Des., 2 (1985), pp. 297–306.

[34] K. Hollig, *Box-spline surfaces*, in Mathematical Methods in Computer Aided Geometric Design, T. Lyche and L. Schumaker, eds., Academic Press, New York, 1989, pp. 385–402.

[35] I. Ihm, *Surface design with implicit algebraic surfaces*, Ph.D. Thesis, Computer Science, Purdue University, 1991.

[36] A. Jones, *Non-rectangular surface patches with curvature continuity*, Computer Aided Design, 20 (1988), pp. 325–335.

[37] C. Lawson and R. Hanson, *Solving Least-Squares Problems*, Prentice-Hall, Englewood Cliffs, NJ, 1974.

[38] E. Lee, *The rational Bézier representation for conics*, in Geometric Modelling: Algorithms and New Trends, G. Farin, ed., Society for Industrial and Applied Mathematics, Philadelphia, 1987, pp. 3–19.

[39] W. Lorensen and H. Cline, *Marching cubes: A high resolution 3D surface construction algorithm*, Computer Graphics, 21 (1987), pp. 163–169.

[40] R. Lorentz, *Uniform bivariate Hermite interpolation*, in Mathematical Methods in Computer Aided Geometric Design, T. Lyche and L. Schumaker, eds., Academic Press, New York, 1989, pp. 435–444.

[41] A. Middleditch and K. Sears, *Blend surfaces for set theoretic volume modeling systems*, Computer Graphics (Proc. SIGGRAPH '85), 19 (1985), pp. 161–170.

[42] D. Moore and J. Warren, *Approximation of dense scattered data using algebraic surfaces*, Proc. of the 24th Hawaii Intl. Conference on System Sciences, Kauai, Hawaii, 1991, pp. 681–690.

[43] J. Morgan and R. Scott, *A nodal basis for C^1 piecewise polynomials of degree $n \geq 5$*, Math. Comp., 29 (1975), pp. 736–740.

[44] G. Nielson, *A transfinite, visually continuous triangular interpolant*, in Geometric Modelling: Algorithms and New Trends, G. Farin, ed., Society for Industrial and Applied Mathematics, Philadelphia, 1987, pp. 235–245.

[45] J. Owen and A. Rockwood, *Blending surfaces in solid geometric modeling*, in Geometric Modelling: Algorithms and New Trends, G. Farin, ed., Society for Industrial and Applied Mathematics, Philadelphia, 1987, pp. 347–366.

[46] K. Paton, *Conic sections in chromosome analysis*, Pattern Recognition, 2 (1970), pp. 39–51.

[47] J. Peters, *Smooth mesh interpolation with cubic patches*, Comput. Aided Des., 22 (1990), pp. 109–120.

[48] B. Piper, *Visually smooth interpolation with triangular Bézier patches*, in Geometric Modelling: Algorithms and New Trends, G. Farin, ed., Society for Industrial and Applied Mathematics, Philadelphia, 1987, pp. 221–233.

[49] M. Powell and M. Sabin, *Piecewise quadratic approximations on triangles*, ACM Trans. Math. Software, 3 (1977), pp. 316–325.

[50] V. Pratt, *Direct least squares fitting of algebraic surfaces*, Computer Graphics, (Proc. SIGGRAPH '87), 21 (1987), pp. 145–152.

[51] J. Rossignac and A. Requicha, *Constant-radius blending in solid modeling*, Computers in Mathematical Engineering, 2 (1984), pp. 655–673.

[52] M. Sabin, *Open questions in the application of multivariate B-splines*, in Mathematical Methods in Computer Aided Geometric Design, T. Lyche and L. Schumaker, eds., Academic Press, New York, 1989, pp. 529–538.

[53] P. Sampson, *Fitting conic sections to very scattered data: An iterative refinement of the Bookstein algorithm*, Computer Graphics and Image Processing, 18 (1982), pp. 97–108.

[54] R. Sarraga, *G^1 interpolation of generally unrestricted cubic Bézier curves*, Comput. Aided Geom. Des., 4 (1987), pp. 23–39.

[55] T. Sederberg, *Piecewise algebraic surface patches*, Comput. Aided Geom. Des., 2 (1985), pp. 53–59.

[56] L. Shirman and C. Sequin, *Local surface interpolation with Bézier patches*, Comput. Aided Geom. Des., 4 (1987), pp. 279–295.

[57] G. Strang, *Piecewise polynomials and the finite element method*, Bull. Amer. Math. Soc., 79 (1973), pp. 1128–1137.

[58] J. Warren, *Blending algebraic surfaces*, ACM Trans. Graphics, 8 (1989), pp. 268–273.

[59] W. Whitely, *A matrix for splines*, J. Approx. Theory, (1989).

[60] J. Woodwark, *Blends in geometric modeling*, in The Mathematics of Surfaces II, R. Martin, ed., Clarendon Press, Oxford, pp. 255–297.

[61] O. Zariski and P. Samuel, *Commutative Algebra (Vol.I, II)*, Springer-Verlag, New York–Berlin, 1958.

Functional Splines for Interpolation, Approximation, and Blending of Curves, Surfaces, and Solids

Josef Hoschek and Erich Hartmann

3.1. Introduction

Most modeling systems are very limited in the complexity of the surfaces that they support. In general these systems use parametric representations of free-formed surfaces like Gordon–Coons surfaces, Bézier surfaces, and B-spline surfaces. In such systems special problems occur if *splines with continuity conditions of higher-order* are required or *blending of surfaces* [11], [12], [14], [20]–[22], [31] or *filling of surface holes* [24], [6], [7], [25] or *approximation of solids* are wanted. We introduce a new method that can be used for interpolation and approximation of curves, surfaces, and solids, as well as for blending and filling of surface holes. This procedure does not use parametric curves or surfaces, but algebraic or transcendental curves and surfaces. This method is a generalization of Liming's classical work [19] on pencil of conics (see also [29], [30]).

Throughout this paper we deal with curves and surfaces described by *implicit* equations. If f and g are two curves (surfaces), then we will consider composite curves (surfaces) of the form

$$F = (1 - \mu)f - \mu g^n = 0.$$

We call f the *base curve* (surface) and g the *transversal curve* (surface).

3.2. G^{n-1}-Functional Splines in the Plane

We want to extend the conic blending method in $\mathbb{R}^2$ introduced by Liming to sufficiently smooth implicit plane curves; further, we give some statements about convexity and apply this method to the interpolation of polygons.

First we introduce two definitions.

DEFINITION 3.2.1. For an implicit curve h we set

$$D^+(h) = \{X \in \mathbb{R}^2 | h(x) \geq 0\}.$$

DEFINITION 3.2.2. Let f and g be two piecewise C^1-continuous curves in $\mathbb{R}^2$ with

53

(i) $|f \cap g| \geq 1$,

(ii) $(f_x(P), f_y(P)) \neq (0,0)$ for all $P \in f \cap g$ (with f_x, f_y as partial derivatives).

Then the curve F defined by the equation

(3.1) $F = (1 - \mu)f - \mu g^n = 0, \qquad 0 < \mu < 1, n \geq 2,$

is called a *plane functional spline* related to the curves f and g. f is the base curve and g the transversal curve of F. n is the *exponent* and μ the *parameter* of F.

$$D(f,g) := D^+(f) \cap D^+(g), \quad D_0(f,g) := D(f,g) \backslash (f \cup g).$$

$\Phi(f,g)$ denotes the family of the plane functional splines related to f and g (μ varies between 0 and 1).

The essential property of a functional spline gives

THEOREM 3.2.1. *Let F be a functional spline curve, f its base curve, g its transversal curve, n its exponent, and $P \in f \cap g$ where f and g are C^n at P. Then at point P the functional spline curve F and the base curve f have contact of order (at least) $n - 1$. If F has exponent $n \geq 3$, then F and f have the same curvature at P.*

Remarks.

1. The proof follows from straightforward calculations and uses Taylor expansion [18].

2. For $n > 3$ the curves F and f even have equal $(n - 3)$th derivatives of the curvature at P (see [26]).

3. Similar results were developed independently in [5].

4. A functional spline curve F with exponent n and C^n-continuous base and transversal curves can be called a G^{n-1} *functional spline curve segment.*

The application of Theorem 3.2.1 to the lines l_1, l_2, l_0 yields that the functional splines

$$F = (1 - \mu)l_1 l_2 - \mu l_0^n = 0 \qquad (0 < \mu < 1, n \geq 3)$$

have curvature 0 at points P_1 and P_2, respectively (Fig. 3.1). Figure 3.2 shows an example with nonlinear base curves and a linear transversal with the equation

$$(1 - \mu)(x^2 + y^2 - r^2)((x - x_0)^2 + y^2 - r^2) - \mu y^3 = 0.$$

The total curve consists of four functional spline segments. The boundary points of the segments are the intersection points of the line and the circles. In these boundary points the spline segments are G^2-continuous and have the same curvature as the given circles.

We will now prove some global geometric properties of functional spline curves.

PROPOSITION 3.2.1. *A functional spline F related to two curves f and g has the property*

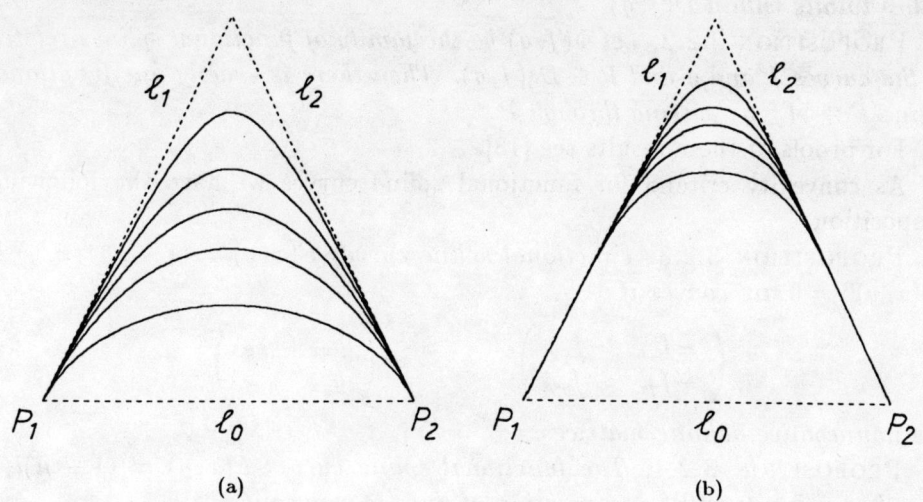

FIG. 3.1. *Functional spline curves of degree* (a) $n = 3$ *and* (b) $n = 8$ *and different values of the parameter* μ.

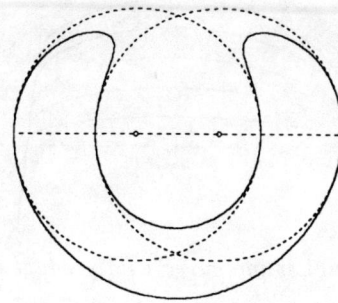

FIG. 3.2. *Functional splines with circular base curves and* $\mu = 0.9$. *Note: To generate the figures a marching algorithm is used (see also* [8], [10]). (1) *Determine in point* X_i *the tangent* t_i *(perhaps an approximation by the secant* $\overline{X_{i-1}X_i}$) *and take, as the first approximation for* X_{i+1}, *a point* Q *on the tangent* t_i (*Fig.* 3.6). (2) *Iteration on a circle through* Q *with center* X_i *yields* X_{i+1}.

$$F \cap f = F \cap g = f \cap g.$$

F lies totally within $D(f, g)$.

PROPOSITION 3.2.2. *Let $\Phi(f, g)$ be the family of functional splines related to the curves f and g and $P \in D_0(f, g)$. Then there is exactly one functional spline $F \in \Phi(f, g)$ passing through P.*

For proofs of these results see [18].

As convexity criteria for functional spline curves we have the following proposition.

PROPOSITION 3.2.3. *Functional spline curves $F(x, y) = (1 - \mu)f(x, y) - \mu g(x, y)^n = 0$ are convex if*

$$\begin{pmatrix} -f_{xx} & -f_{xy} \\ -f_{yx} & -f_{yy} \end{pmatrix}, \qquad \begin{pmatrix} g_{xx} & g_{xy} \\ g_{yx} & g_{yy} \end{pmatrix}$$

are nonnegative definite matrices.

PROPOSITION 3.2.4. *The functional spline curves $F(x, y) = (1 - \mu)l_1 \cdot l_2 \cdots l_k - \mu l_0^n = 0$ with l_j as equation of lines is convex if $n \geq k$.*

The proofs follow by calculating the curvature of the functional spline curves or using criteria (*) in Definition 3.3.3, which is necessary and sufficient for convex curves (see [4]).

Figure 3.3 shows that a functional spline curve can also be convex for $n \leq k$. We have chosen ten lines $l_1, \cdots, l_{10}$ and as an exponent $n = 3$.

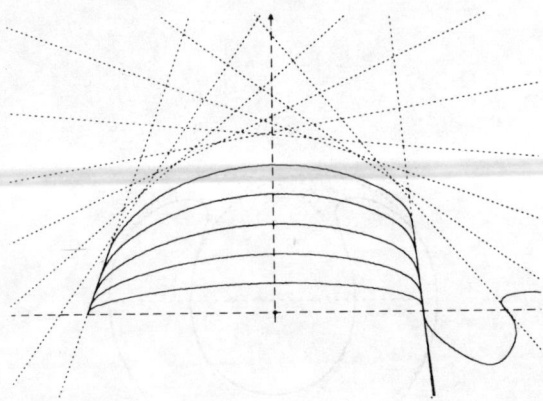

FIG. 3.3. *Convex functional spline curve with product of ten lines as base curves.*

As an application of the last results to a convex polygon we have the following theorem.

THEOREM 3.2.2. *For any plane convex polygon there exists an infinite number of convex G^2-interpolation functional splines containing the vertices. Such an interpolation curve consists of implicit segments of degree (at least) three.*

From Proposition 3.2.4 and Theorem 3.2.2 we get the following construction. The set of functional splines

$$(1 - \mu_i)l_i l_{i+1} - \mu_i l_{0i}^3 = 0 \quad (0 < \mu_i < 1), \quad i = 1, 2, \cdots$$

are cubic curves that interpolate the given convex polygon. These spline segments are G^2-continuous.

In Fig. 3.4 we show an example. A hexagon is interpolated with the help of another circumscribed hexagon. The edges of the inner hexagon are the transversal curves; the edges of the circumscribed hexagon are the base curves. The exponent $n = 3$ is chosen; therefore the spline segments are G^2-continuous.

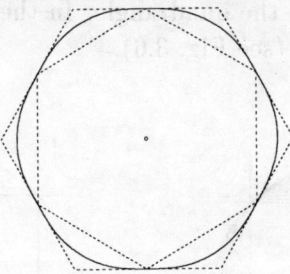

FIG. 3.4. *Interpolation of the vertices of a hexagon with G^2-functional spline segments.*

The disadvantage of Theorem 3.2.2 is that at every vertex of the polygon the curvature is equal to zero. In order to avoid this defect, one can, for instance, choose convenient quadratic curves c_i instead of the lines l_i. These curves c_i should be expressed in a local coordinate system by an explicit form in order to guarantee convexity by Proposition 4 in [18]. A higher order of smoothness can be obtained by raising the exponents of the functional splines. Figure 3.5 shows the interpolation of a square by functional splines with (a) lines and (b) a circle as base curves.

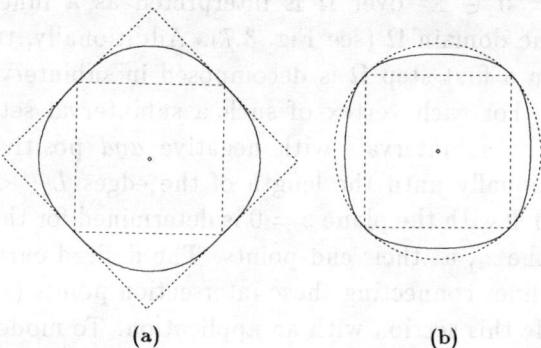

(a)   (b)

FIG. 3.5. *Interpolation of a square by four G^2-functional spline segments.* (a) *Base curves: lines* $(n = 4, \mu = \frac{2}{3})$. (b) *Base curve: circle* $(n = 4, \mu = 0,95)$.

A direct working method for interpolating polygons with functional curves was introduced in [28]. The interpolation of a quadrangle with the help of a circle follows, for instance, with the interpolator

$$f = K - \lambda Q = 0 \qquad (\lambda \in \mathbb{R}^+)$$

with K as a circle

$$K(x, y) = x^2 + y^2 - 2a^2 = 0$$

and Q as the product of the equations of the edges of the quadrangle

$$Q = (x + a)(x - a)(y + a)(y - a) = 0.$$

For $\lambda \to 0$ the set of curves f converges to the circle K; for $\lambda \to +\infty$ the set of curves f converges to the quadrangle. In the vertices the interpolator is C^1-continuous to the circle (see Fig. 3.6).

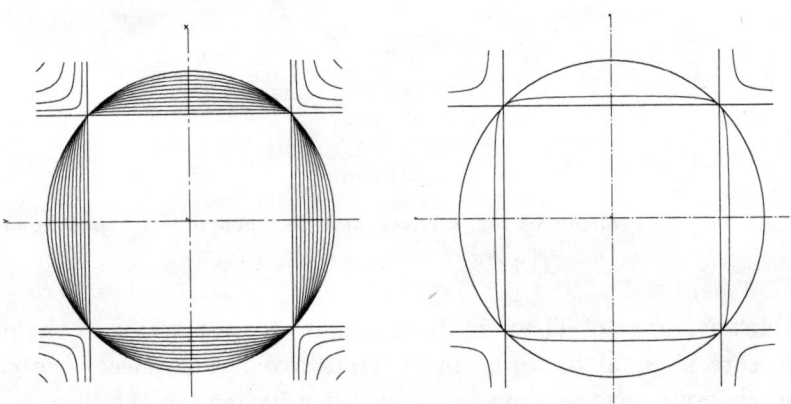

FIG. 3.6. *Interpolation of the vertices of a quadrangle with the help of a circle. The left-hand side contains some curves out of the set of curves; the right-hand side curve shows one interpolating curve.*

Also in [28], another method for curve representation was used. The function $f(x, y) = 0 \in \mathbb{R}^2$ over Ω is interpreted as a function in $\mathbb{R}^3 \Phi :=$ $z - f(x, y)$ over the domain Ω (see Fig. 3.7). Additionally, two error bounds ϵ_1, ϵ_2 are given. In a first step Ω is decomposed in subintervals with lengths of edges $l_1 < \epsilon_1$. For each vertex of such a subinterval set $z_{ij} = f(x_i, x_j)$ is determined. The subintervals with negative *and* positive z_{ij} values are decomposed additionally until the length of the edges $L_{2k} < \epsilon_2$ is obtained. The intersection of Φ with the plane $z = 0$ is determined for those edges having different signs of the z_{ij} at their end points. The desired curve is then drawn as the polygon of lines connecting these intersection points (see Fig. 3.8).

Let us conclude this section with an application. To model a projection of a human face, we described the first approximation by straight lines (see Fig. 3.9); afterwards we smooth these polygons with different parameters μ (see Fig. 3.10).

3.3. G^{n-1}-Functional Spline Surfaces

In this section we extend the idea of functional spline curves to functional spline surfaces and show that in $\mathbb{R}^3$ we get analogous results. First we extend Definitions 3.2.1 and 3.2.2.

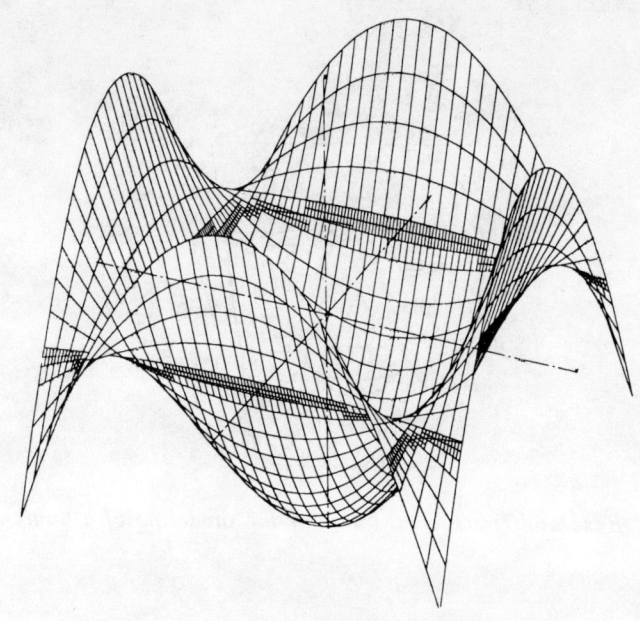

FIG. 3.7. *Interpretation of the interpolant from Fig. 3.6 in* $\mathbb{R}^3$.

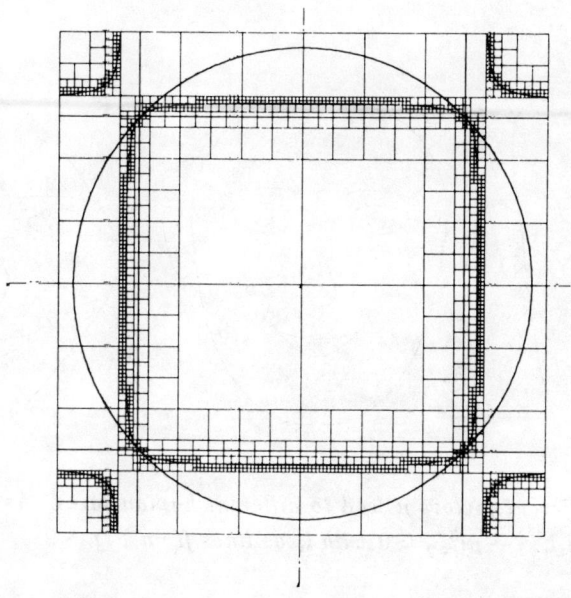

FIG. 3.8. *Decomposition of* Ω *for interpolation of a quadrangle with the help of the curve from Fig. 3.6 (right).*

FIG. 3.9. *Base and transversal lines for the modeling of a human face.*

FIG. 3.10. *Different factors μ lead to different human faces. As an interpolator was chosen $(1-\mu)l_i l_{i+1} - \mu l_{i+2}^3 = 0$ with l_i as lines from Fig. 3.9.*

DEFINITION 3.3.1. For an implicit surface h we set $D^+(h) = \{X \in \mathbb{R}^3 | h(X) \geq 0\}$.

DEFINITION 3.3.2. Let f and g be two piecewise C^1-continuous surfaces. (i) $f \cap g$ consists of a curve Γ. (ii) $(f_x(P), f_y(P), f_z(P)) \neq (0,0,0)$ for "nearly" all $P \in \Gamma$ (only a finite number of exceptions). Then the surface F defined by the equation

$$F = (1 - \mu)f - \mu g^n = 0, \quad 0 < \mu < 1, \quad n \geq 2,$$

is called a *functional spline surface* related to the surfaces f and g. f is the base surface and g the transversal surface of F. n is the exponent and μ the parameter of F.

$$D(f,g) := D^+(f) \cap D^+(g), \qquad D_0(f,g) := D(f,g) \backslash (f \cup g).$$

$D(f,g)$ denotes the family of the functional splines related to f and g (μ varies between 0 and 1).

Now we can introduce the following theorem.

THEOREM 3.3.1. *Let F be a functional spline surface, f its base surface, and g its transversal surface, n its exponent, and $P \in f \cap g$ where f and g are C^n at P. Then at point P the functional spline F and its base surface f have contact of order (at least) $n - 1$. If F has an exponent ≥ 3, then F and f even have the same Dupinian indicatrix at P.*

Remarks.

1. For the proof see [18].

2. From Theorem 3.3.1 follows that for $n \geq 3$, F and f have (except a finite number of points) the same Gaussian curvature along Γ (see [1]–[3], [17], [27]).

3. A functional spline surface F with exponent n and C^n-continuous base and transversal surfaces is called a G^{n-1} *functional spline surface patch*.

Figure 3.11 contains the interpolation of a circle c with radius r by a functional spline: the base surface is the upper half sphere containing c.

$$(1 - \mu)(x^2 + y^2 + z^2 - r^2) - \mu z^n = 0 \quad \text{and} \quad n = 3, 8.$$

Additionally, we establish a G^2-interpolation of two spatial curves with a common segment. Let Γ_1, Γ_2 be two sufficiently smooth curves with the common segment Γ (Fig. 3.12). Take (if possible) convenient surfaces f_1, f_2 containing Γ_1 (resp. Γ_2) and a surface f containing Γ. Further we need transversal surfaces g_1, g_2 passing $\Gamma_1 \cup \Gamma$ (resp. $\Gamma_2 \cup \Gamma$). Then the functional splines

$$F_1 = (1 - \mu)f_1 f - \mu g_1^n = 0, \qquad 0 < \mu < 1, \qquad n \geq 3,$$
$$F_2 = (1 - \lambda)f_2 f - \lambda g_2^m = 0, \qquad 0 < \lambda < 1, \qquad m \geq 3,$$

interpolate $\Gamma_1 \cup \Gamma$ (resp. $\Gamma_2 \cup \Gamma$) and have (because of Theorem 3.3.1) the same tangent planes and Dupinian indicatrices along the common curve Γ. Thus $F_1 \cup F_2$ is a G^2-interpolation surface containing $\Gamma_1 \cup \Gamma_2 \cup \Gamma$.

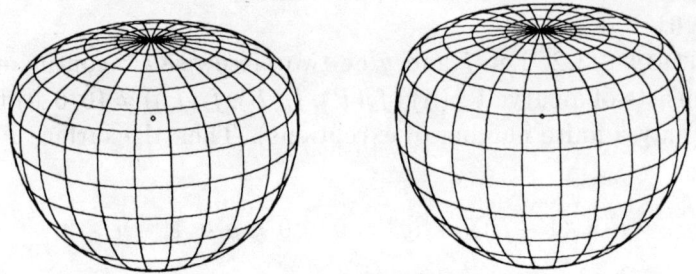

FIG. 3.11. G^2-interpolation of a circle. Base surface is a sphere; transversal surface is the plane through the midpoint of the sphere and the circle.

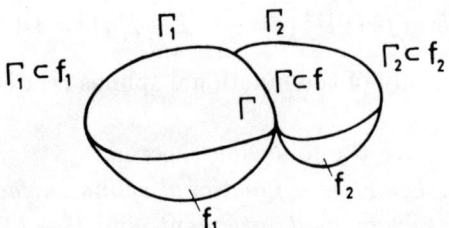

FIG. 3.12.   Interpolation of neighboring curves.

Figure 3.13 contains the following example of the last result. Γ_1 and Γ_2 are the halves of the circle $c : x^2 + y^2 = r^2$ with $x \geq 0$ (resp. $x \leq 0$); Γ is a diameter of the circle. f_1, f_2 are quarters of the sphere containing the circle c; f is the plane $x = 0$; and g_1 and g_2 are the plane $z = 0$. Thus, we get the functional splines

$$F_1 = (1 - \mu)(r^2 - x^2 - y^2 - z^2)x - \mu z^3 = 0 \qquad \text{(left part)},$$
$$F_2 = (1 - \lambda)(r^2 - x^2 - y^2 - z^2)(-x) - \lambda(-z)^3 = 0 \qquad \text{(right part)}.$$

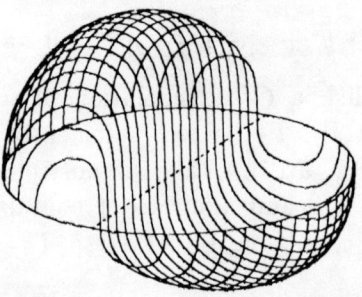

FIG. 3.13.   Interpolation of a circle and a line with help of the sphere.

3.3.1. Interpolation and Approximation with G^{n-1}-Functional Spline Surface Patches.   First we introduce convexity theorems and start with a definition.

DEFINITION 3.3.3. If an implicit surface $f(x_1, \cdots, x_m) = 0 \subset \mathbb{R}^m$ satisfies

$$(*) \qquad \begin{vmatrix} f_{11} & f_{12} & \cdots & f_{1k} & f_1 \\ f_{21} & f_{22} & \cdots & f_{2k} & f_2 \\ \vdots & \vdots & & \vdots & \vdots \\ f_{k1} & f_{k2} & \cdots & f_{kk} & f_k \\ f_1 & f_2 & \cdots & f_k & 0 \end{vmatrix} \leq 0 \qquad (k = 1, \cdots, m)$$

with

$$f_j = \frac{\partial f}{\partial x_j}, \qquad f_{ij} = \frac{\partial^2 f}{\partial x_i \partial x_j},$$

then we call the surface convex at the point $x = (x_1, \cdots, x_m)$.

Remark. For $m = 2$ (see also Proposition 3.2.3) and $m = 3$ this condition is necessary and sufficient for convexity [17].

Additionally, we need the following proposition.

PROPOSITION 3.3.1. *Let $A = (a_{ij})$ be an $n \times n$ matrix. If A is nonnegative definite, then for any n vector $(b_1 \cdots b_n) \in \mathbb{R}^n$ we have*

$$(*) \qquad D_{n+1} = \begin{vmatrix} a_{11} & a_{12} & \cdots & a_{1n} & b_1 \\ \vdots & \vdots & \vdots & \vdots & \vdots \\ a_{n1} & a_{n2} & \cdots & a_{nn} & b_n \\ b_1 & \cdots & & b_n & 0 \end{vmatrix} \leq 0.$$

The proof follows from straightforward calculations (see also [23]). Thus we can prove the following.

THEOREM 3.3.2. *The functional spline*

$$F = (1 - \mu)f - \mu g^n = 0$$

is convex in $\Omega(f, g)$ if $f, g \in C^2(\Omega)$ and M_f, M_g are nonnegative definite matrices; here $\Omega(f, g) = \{x \in \mathbb{R}^m | f(x) \geq 0, g(x) \geq 0\}$ and

$$M_h = \begin{pmatrix} h_{11} & h_{12} & \cdots & h_{1m} \\ h_{21} & h_{22} & \cdots & h_{2m} \\ \vdots & \vdots & \cdots & \vdots \\ h_{m1} & h_{m2} & \cdots & h_{mm} \end{pmatrix}$$

with $h_{ij} = \frac{\partial^2 f}{\partial x_i \partial x_j}$.

From Theorem 3.3.2 we get for surfaces in $\mathbb{R}^3$ the following propositions.

PROPOSITION 3.3.2. *The functional surface patch $F = (1 - \mu)f - \mu g^n = 0$ is convex in $\Omega(f, g) = \{(x, y, z) \in \mathbb{R}^3 : f(x, y, z) \geq 0, g(x, y, z) \geq 0\}$ if M_{-f}, M_g are nonnegative definite matrices with*

$$M_h = \begin{pmatrix} h_{11} & h_{12} & h_{13} \\ h_{21} & h_{22} & h_{23} \\ h_{31} & h_{32} & h_{33} \end{pmatrix}, \qquad h_{ij} = \frac{\partial^2 h}{\partial x_i \partial x_j}.$$

PROPOSITION 3.3.3. *The functional spline surface*

$$F = (1 - \mu)l_1 \cdot l_2 \cdots l_k - \mu l_0^n = 0$$

with $l_j = 0$ as equation of planes is convex if $n \geq k$.

Remark. For the proof see [4].

We will use these results for interpolation and approximation of solids. First we have the following theorem.

THEOREM 3.3.3. *For any convex polyhedron in $\mathbb{R}^3$ there is an infinite number of convex interpolation surfaces that pass through all edges and are C^2 everywhere except at the vertices. To a face π_i of the polyhedron with N_i edges the corresponding part S_i of the interpolation surface has degree $\geq N_i$.*

From Proposition 3.3.3 and Theorem 3.3.3 follows the construction.

Construction. Through each edge of the polyhedron we construct a new plane (different from all planes containing faces) so that the polyhedron lies on one side of the plane (see Fig. 3.14). If p_0 is the plane that contains the face π_i and if $p_1, p_2, \cdots$ are the "new" planes corresponding to the face π_i, we interpolate this face by a functional spline

$$(*) \qquad F = (1 - \mu)p_1 p_2 \cdots - \mu p_0^3 = 0 \qquad (0 < \mu < 1)$$

The set of all functional spline patches (corresponding to the faces) make up a piecewise interpolation spline surface of the polyhedron.

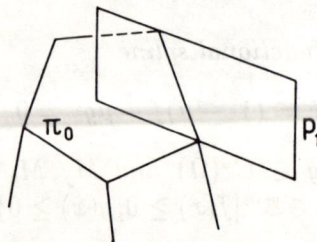

FIG. 3.14. *Interpolation of the face of a polyhedron.*

Although the above surface is joining G^2-continuously, there are flat points along the edges. In order to avoid this defect, we can use other base surfaces along the edges, for example, quadrics.

The following figures demonstrate the effectiveness of the introduced method. Figures 3.15–3.22 show the interpolation of a square, a cube, and a dodecagon. Figure 3.15 describes the interpolation of a square with the function

$$(1 - \mu)(z + x - a)(z - x - a)(z + y - a)(z - y - a) - \mu z^3 = 0,$$

where planes are used as base surfaces. In Fig. 3.16(a) the surface in Fig. 3.15 is transformed to the faces of a cube. Along the edges the interpolation surface has flat points (G^2-continuous to a plane!). This shortcoming is avoided in Fig. 3.16(b). We use cylinders as base surfaces

$$(1-\mu)\left(z^2+x^2-\frac{a^2}{2}\right)\left(z^2+y^2-\frac{a^2}{2}\right)-\mu\left(z-\frac{a}{2}\right)^3=0.$$

The interpolation surfaces in Figs. 3.16(a), (b) have singularities at the vertices of the cube. The singularity at the vertex $V(\frac{a}{2},\frac{a}{2},\frac{a}{2})$ can be removed if we cut out a region Φ about V with the plane $\Pi : x+y+z-a=0$ and replace Φ by the surface with the equation

$$F(x,y,z)=(1-\lambda)f_1(x,y,z)f_2(x,y,z)f_3(x,y,z)-\lambda(x+y+z-a)^3=0,$$
$$0<\lambda<1,$$

with f_i as the corresponding face surfaces of the interpolation surface in the vertex V (see Fig. 3.16(c)). This *approximating* functional spline surface is totally G^2-continuous.

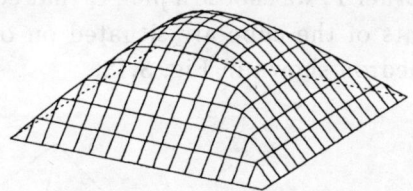

FIG. 3.15. *Interpolation of a square. Beside the vertices the interpolating surface is G^2-continuous.*

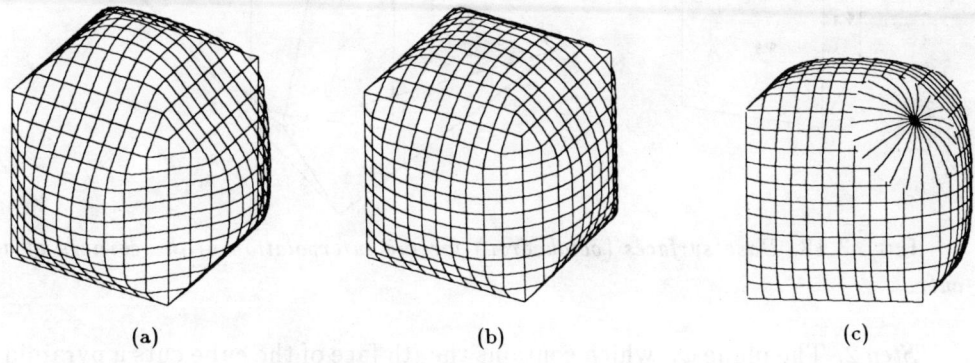

(a)   (b)   (c)

FIG. 3.16. (a) *Interpolation of a cube with the help of planes (observe the flat points at the edges). (b) Interpolation of a cube with help of cylinders. (c) Removal of a singularity from the example in Fig. (a).*

Figure 3.17 contains the interpolation of a dodecagon by the functional spline

$$(1-\mu)l_1\cdots l_{12}-\mu z^3=0 \qquad \text{(with } l_i \text{ as suitable chosen planes).}$$

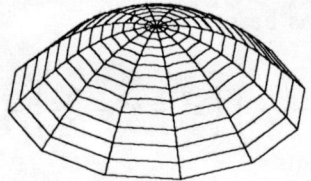

FIG. 3.17. *Interpolation of a dodecagon (beside the vertices the interpolation surface is regular).*

The interpolation surface is smooth. The corners, which appear in the figures, follow from the projection of the surfaces curves.

In Figs. 3.15 and 3.16 the interpolation of a cube is developed while the interpolation surface contains the edges of the cube. Now we will describe a procedure for constructing a family of G-Interpolation surfaces which pass *only* the corners of a cube. The method used can be generalized to arbitrary convex polyhedrons.

Step 1. For each corner P_i we choose a plan ϵ_i that contains P_i and supports the cube, i.e., all points of the cube are situated on one side of ϵ_i. Here we take the regular octahedron shown in Fig. 3.18.

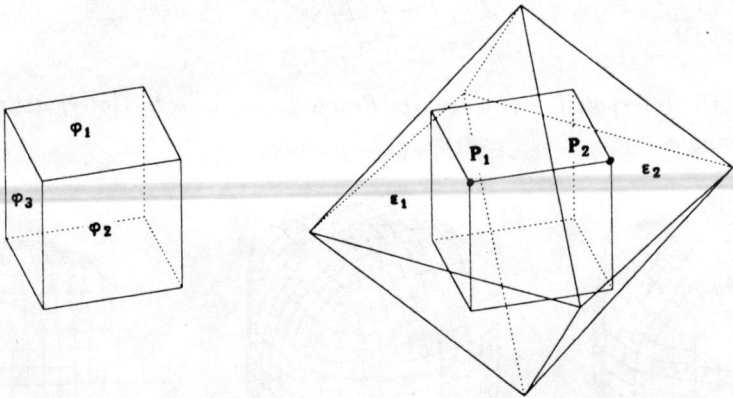

FIG. 3.18. *Base surfaces (octahedron) for the interpolation of the corners of a cube.*

Step 2. The plane φ_i, which contains the ith face of the cube cuts a pyramid out of the octahedron (Fig. 3.19). With the method introduced by Theorem 3.3.3, we construct upon φ_i a G^2-interpolation surface Φ_i that contains the base quadrangle of the pyramid and the corners of the ith face of the cube. If we replace the ith pyramid of the octahedron by Φ_i, we get a convex surface which is the boundary of a convex body B_i. In the same way we round all the corners of the octahedron. If we intersect all the convex bodies B_i, $i = 1, \cdots, 6$, we get a convex body whose boundary Φ consists of parts of the surfaces Φ_i, $i = 1, \cdots, 6$, as boundary patches. (Figure 3.19 shows three neighboring patches.) The surface Φ has singular points, namely the intersection curves of neighboring surfaces Φ_i.

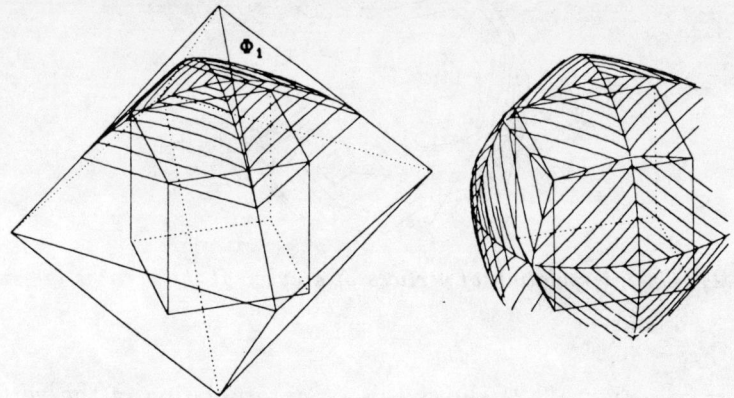

FIG. 3.19.   *Interpolation surfaces of the corners of the cube.*

Step 3.   We intersect every plane φ_i with the surface Φ and get a set of auxiliary curves on Φ (Fig. 3.20).

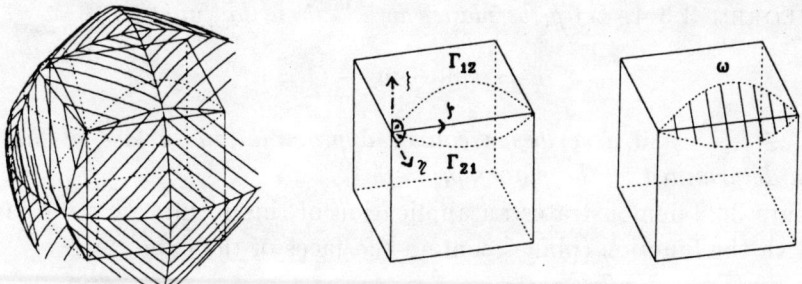

FIG. 3.20.   *Auxiliary curves of $\varphi_i \cap \Phi_i$.*

Let Γ_{12}, Γ_{21} be the auxiliary curves belonging to the planes φ_1, φ_2 and let e_{12} be the edge between faces 1 and 2 of the cube. We introduce a local ξ-η-ζ-coordinate system such that e_{12} lies on the ζ-axis, the ξ-η-plane is orthogonal to e_{12}, φ_1 contains the ξ-axis, and φ_2 contains the η-axis (Fig. 3.20). Then Γ_{12} and Γ_{21} can be described by equations $\xi = \gamma_1(\zeta)$ and $\eta = \gamma_2(\zeta)$, respectively. We take the ruled surface ω with the equation

$$g(\xi, \eta, \zeta) = \xi\gamma_2(\zeta) + \eta\gamma_1(\zeta) - \gamma_1(\zeta)\gamma_2(\zeta) = 0$$

as the base surface of functional spline Ψ:

$$F = (1 - \lambda)f_1 f_2 - \lambda g^n = 0, \quad 0 < \lambda < 1, \quad n > 2,$$

where $f_1(\xi, \eta, \zeta) = 0$ and $f_2(\xi, \eta, \zeta) = 0$ are the equations of Φ_1 and Φ_2, respectively. The functions γ_1, γ_2 are convex, even in the general case. In the case of the cube we can choose the parameters of the functional splines f_1, f_2

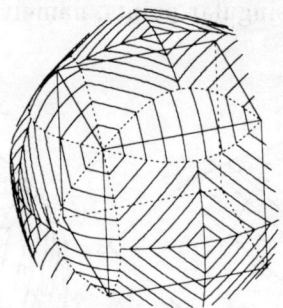

FIG. 3.21. G^2-*Interpolation of vertices of a cube: at the vertices the surface has* *flat points.*

so that $\gamma_1 = \gamma_2$. Figure 3.21 shows this G^2-interpolation of the vertices of a cube.

We can also construct pure approximation surfaces of a polyhedron with the help of the product of the planes p_i of the faces of the polyhedron. By straightforward calculation with help of the convexity criterion in Definition 3.3.3 we can prove the following theorem.

THEOREM 3.3.4. *Let p_i be planes in $\mathbb{R}^3$, then the function*

$$F = p_0 p_1 \cdots p_n - C = 0$$

with $C \in \mathbb{R}, C \neq 0$ describes a convex approximation surface of the convex domain determined by the planes p_i.

Figure 3.22 demonstrates an application of this result: we approximate a cube with the function (planes contain the faces of the cube)

$$F = (a - x)(a + x)(a - y)(a + y)(a - z)(a + z) \cdot a^{-6} - C = 0$$

with $2a$ as lengths of the edge of the cube and $C = 0.1$.

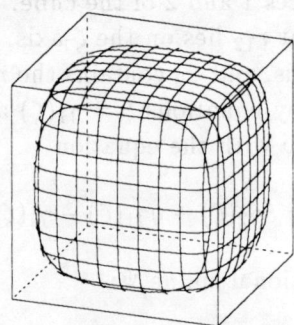

FIG. 3.22. *Approximation of a cube.*

3.3.2. Rounding of solids. Now we show an application of the functional splines on the problem of the rounding of the edges and the vertices of solids.

We start with the cube (Fig. 3.23) $(0,0,0)$, $(a,0,0)$, $(0,a,0)$, $(a,a,0)$, $(0,0,a)$, $(a,0,a)$, $(0,a,a)$, (a,a,a) and consider the edges e_1, e_2, e_3 of the faces φ_1, φ_2, φ_3, respectively, and the corner V. We smooth the edges in the marked areas (dotted lines in Fig. 3.15) by the functional spline surfaces:

$$f_1(x,y,z) = (1 - \mu)(a - z)(a - y) - \mu(z + y - 3a/2)^3 = 0, \qquad a/2 \le y, z \le a,$$
$$f_2(x,y,z) = (1 - \mu)(a - z)(a - x) - \mu(z + x - 3a/2)^3 = 0, \qquad a/2 \le x, z \le a,$$
$$f_3(x,y,z) = (1 - \mu)(a - x)(a - y) - \mu(x + y - 3a/2)^3 = 0, \qquad a/2 \le x, y \le a,$$
$$0 < \mu < 1.$$

The cylindrical (but not quadric) surfaces f_1, f_2, f_3 intersect in curves c_1, c_2, c_3, which meet in the point W. f_1 has G^2 connection to the plane $\epsilon_2 : y = a$ (resp. $\epsilon_3 : z = a$) that contains the face φ_2 (resp. φ_3) of the cube. Analogous statements hold for f_2 and f_3.

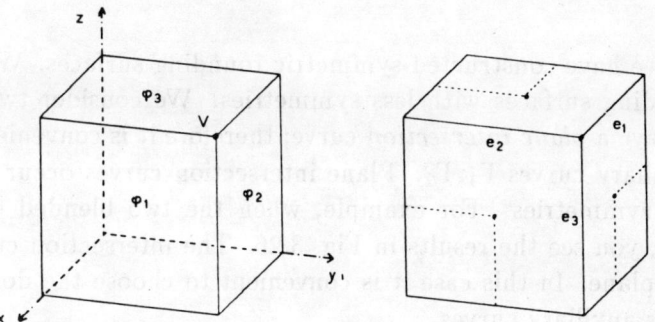

FIG. 3.23.   *Corner V of a cube that should be rounded.*

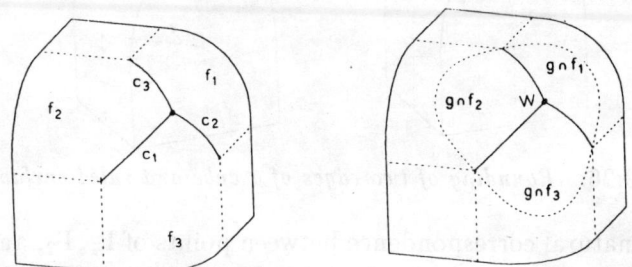

FIG. 3.24.   *Intersection curves of the rounding cylinders.*

To smooth the corner W we use the functional spline surface

$$F = (1 - \lambda)f_1 f_2 f_3 - \lambda g^3 = 0, \qquad 0 < \lambda < 1,$$

where g is the plane $g(x,y,z) = x + y + z - 2a = 0$ passing the points $(a, \frac{a}{2}, \frac{a}{2})$, $(\frac{a}{2}, a, \frac{a}{2})$, $(\frac{a}{2}, \frac{a}{2}, a)$. F has G^2 connections to the surfaces f_1, f_2, f_3. In order to make the surface F visible, we draw curves that are sections with planes containing the diagonal of the cube. Figure 3.25 below shows solutions for various μ and λ.

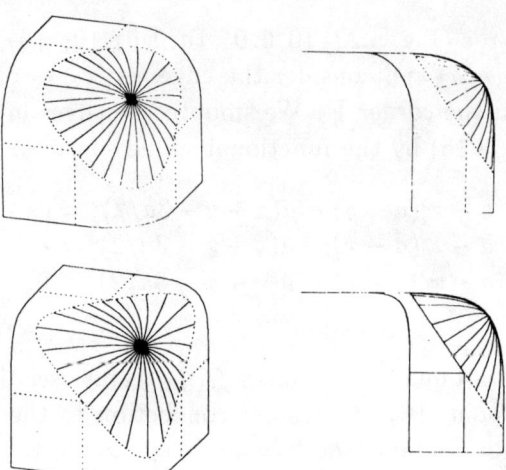

FIG. 3.25. *Rounding surfaces of a vertex of the cube.*

Until now we have constructed symmetric rounding surfaces. We can also introduce rounding surfaces with less symmetries. We consider two surfaces Φ_1, Φ_2 which have a *plane intersection* curve; therefore it is convenient to take also plane auxiliary curves Γ_1, Γ_2. Plane intersection curves occur often and are caused by symmetries. For example, when the two blended edges of a cube are taken, you see the results in Fig. 3.26. The intersection curve Γ lies obviously in a plane. In this case it is convenient to choose the dotted plane curves Γ_1, Γ_2 as auxiliary curves.

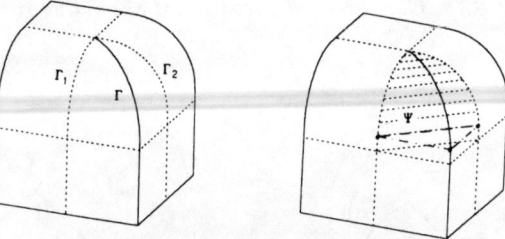

FIG. 3.26. *Rounding of two edges of a cube and ruled surface Ψ.*

There is a natural correspondence between points of Γ_1, Γ_2, and Γ with the same z-coordinates. Corresponding points of Γ_1, Γ_2 are connected by lines and yield the ruled surface Ψ. With a convenient coordinate system (with origin in the hidden corner of Fig. 3.26), the blending surfaces of the two horizontal edges of the suitcase corner have equations

$$f_1(x, y, z) = (1 - \mu)(a - x)(a - z) - \mu \left(z + x - \tfrac{3}{2}a \right)^n = 0$$
$$f_2(x, y, z) = (1 - \mu)(a - y)(a - z) - \mu \left(z + y - \tfrac{3}{2}a \right)^n = 0 \qquad n > 2.$$

Let $x = \varphi(z)$ and $y = \varphi(z)$ be the explicit forms of the equations $f_1 = 0$ and $f_2 = 0$, respectively. Then the ruled surface Ψ (Fig. 3.26) can be described by the equation

$$g(x, y, z) = x + y - \varphi(z) - \frac{a}{2} = 0,$$

and the blending surface Θ (Fig. 3.27) that rounds the suitcase corner has the equation

$$F(x, y, z) = (1 - \lambda)(x - \varphi(z))(y - \varphi(z)) - \lambda \left(x + y - \varphi(z) - \frac{a}{2} \right)^n = 0,$$

$$0 < \lambda < 1, n \geq 3.$$

The surface $F = 0$ joins Φ_1 and Φ_2 G^2-continuously along Γ_1 and Γ_2. Sections of Θ with horizontal planes are used to display the transition surface in Fig. 3.27. If auxiliary curves as shown in Fig. 3.27 are chosen, "toroidal" solutions are obtained.

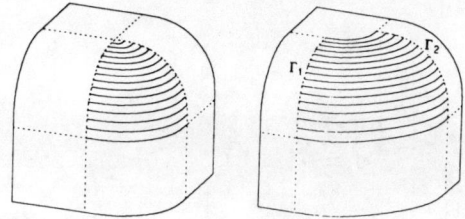

FIG. 3.27. G^2-blending of a suitcase corner.

The same idea (plane auxiliary curves) can be used in the following examples with nonconvex vertices (Fig. 3.28). The equations of the ruled surfaces and the blending surfaces can be evaluated in the same manner shown above. We can also find a rounding surface for a 3-beam corner, a problem which sometimes is called "filling of holes" (see Fig. 3.31).

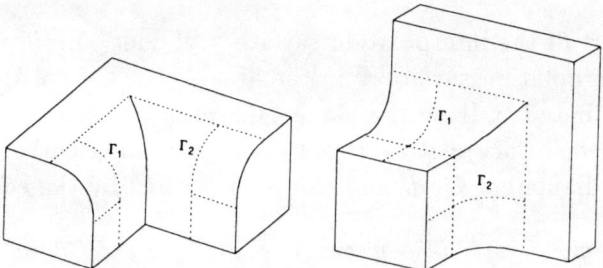

FIG. 3.28.   Toroidal solution for the 2-beam corner.

First we introduce a coordinate system and give some specifications. We choose the coordinate system so that the corner $V_0 = (0, 0, 0)$ and the planes through V_0 are described by the equations $x = 0$, $y = 0$, $z = 0$. We denote the edges that are concurrent in V_0 by e_1, e_2, e_3, e_4, e_5, e_6. The area of interpolation is marked by dotted lines (see Fig. 3.30(a)). We notice that the points V_1, V_2, V_3, V_4, V_5, V_6 are contained in the plane Π: $x + y + z = 0$. This plane will play an important role. We will construct an interpolation surface Φ that passes through the diameters of the hexagon $V_1, \cdots, V_6$ and that has Π

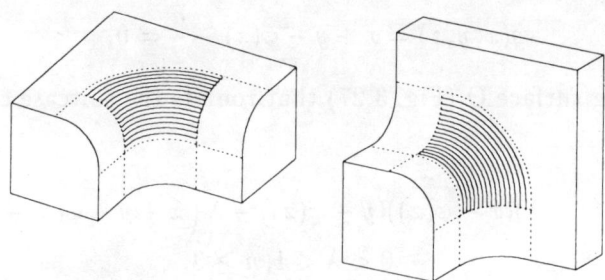

FIG. 3.29. *Toroidal solution for the beam-plane corner.*

as tangent plane in the points of the diameters. Thus the interpolation surface Φ will be determined if we define it in the marked area Ω_1 (Fig. 3.30(b)).

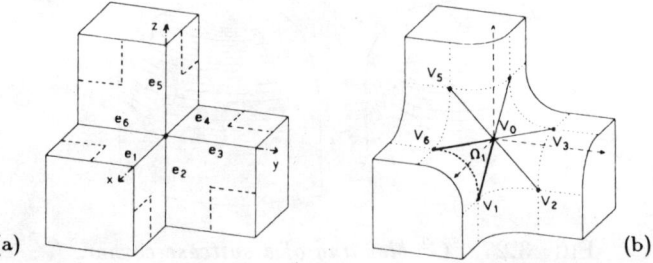

FIG. 3.30. (a) *3-beam corner.* (b) *Rounding of the edges of the 3-beam corner.*

To smooth the *edge* e_1 in the area $x \geq r_c$, we use the method given above for the edges of a cube. Here we get the functional spline:

$$f_1(x, y, z) = (1 - \mu)yz - \mu(y + z + r_c)^3 = 0, \qquad 0 < \mu < 1.$$

We call this part of the interpolation surface "cylinder" Σ.

For the interpolation surface Φ in the area $0 \leq x \leq r_c$ of Ω_1 we construct a functional spline. Let Π be the plane $f_2(x, y, z) = x + y + z = 0$ and Γ be the "cone" (ruled surface) with vertex V_0 and generating curve $f_1 = 0$, $x = r_c$. Γ contains the diameters d_1, d_6 and can be described by the equation

$$g(x, y, z) = (1 - \mu)xyz/r_c - \mu(x + y + z)^3 = 0,$$

where μ is the parameter of f_1. The functional spline surface

$$F = (1 - \lambda)f_1 f_2 - \lambda g^3 = 0, \qquad 0 < \lambda < 1$$

has the demanded properties (contact of order two with Σ and Π).

The following mappings give us the solution in the other areas $\Omega_2, \cdots$.

$$(x, y, z) \rightarrow (-x, -y, -z)$$
$$\rightarrow (\pm y, \pm x, \pm z)$$
$$\rightarrow (\pm y, \pm z, \pm x).$$

The functional splines in Fig. 3.31(a) have the same parameter μ as in the previous figures. Figure 3.31(b) shows solutions for different μ and λ. The curves in these figures are plane sections of the interpolation surfaces.

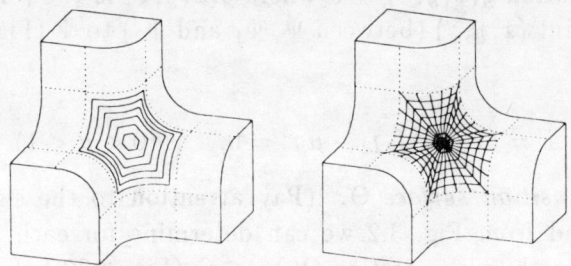

FIG. 3.31. *Different solutions for the rounding of a 3-beam corner.*

3.3.3. Blending with G^2-Functional Splines. We developed an algorithm to blend two implicit surfaces using plane G^2-functional splines that join the given base surfaces G^2-continuously.

Let Φ_1 and Φ_2 be two intersecting surfaces with equations $f_1 = 0$, $f_2 = 0$, respectively, and intersection curve Γ. In order to avoid essential difficulties while introducing the method, some restrictions are imposed:

(i) f_1 and f_2 are differentiable in the area of consideration.

(ii) There is no point in $\Phi_1 \cap \Phi_2$ where the tangent planes coincide.

The idea of the current blending method is to describe the transition surface Θ by plane sections with planes normal to the intersection curve Γ. Thus we get the following.

Step 1. *Determination of points of the auxiliary curves Γ_1 and Γ_2.* With the aid of a "rolling sphere" Σ (radius r_0) (see [13]) to each point X_i, two points $P_1 \in \Gamma_1$ and $P_2 \in \Gamma_2$ are determined. Let Π_1, Π_2 be the tangent planes of the surfaces Φ_1, Φ_2, and Ω the plane containing X_i, which is orthogonal to the intersection curve Γ. The "rolling sphere" has its midpoint in the plane Ω and contacts the planes Π_1, Π_2 in the points Q_1, Q_2 (Fig. 3.32(a)). Let γ be the circle through the point Q_1 (and Q_2) with midpoint X_i. Then P_1 and P_2 are the intersection points $\gamma \cap \Phi_1$, $\gamma \cap \Phi_2$, respectively (Fig. 3.32).

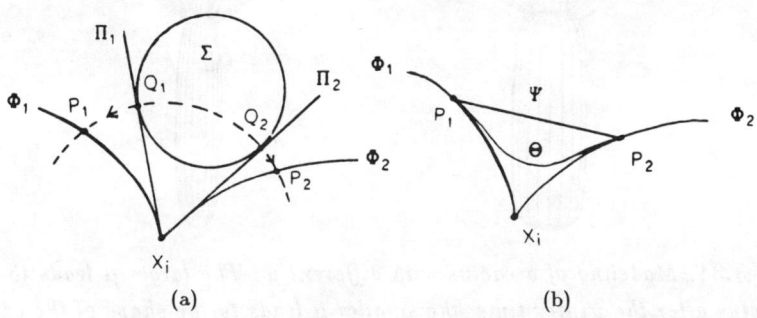

(a)   (b)

FIG. 3.32. (a) *Determination of P_1, P_2.* (b) *Ruled surface Ψ and functional spline Θ.*

Step 2. Determination of the transition surface Θ *by its plane sections with planes orthogonal to the intersection curve* Γ. We use the denotations of Step 1. If the ruled surface Ψ is described in plane sections with planes orthogonal to Γ by the equation $g(x, y, z) = 0$ where $g(x, y, z)$ is the (oriented, normal) distance of a point (x, y, z) (between Ψ, Φ_1 and Φ_2) to Ψ (Fig. 3.32(b)), then the equation

$$F = (1 - \mu)f_1 f_2 - \mu g^3 = 0, \qquad (0 < \mu < 1)$$

represents a *transition surface* Θ. (Pay attention to the signs!) With the marching method from Fig. 3.2 we can determine for each point X_i of the intersection curve the plane section $\Omega(X_i) \cap \Theta$ (Fig. 3.32(b)).

Figure 3.33 shows applications of this blending method.

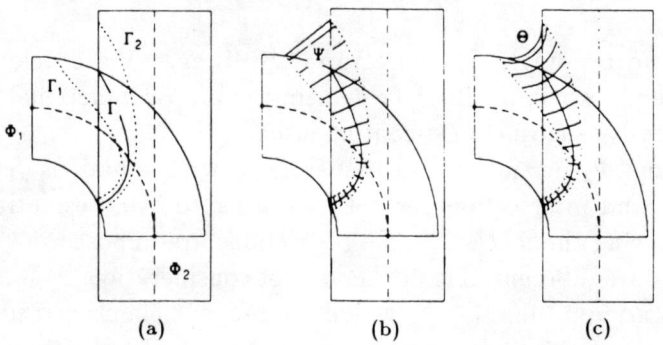

FIG. 3.33. (a) *Auxiliary curves* Γ_1, Γ_2. (b) *Ruled surface* Ψ. (c) *Transition surface* Θ.

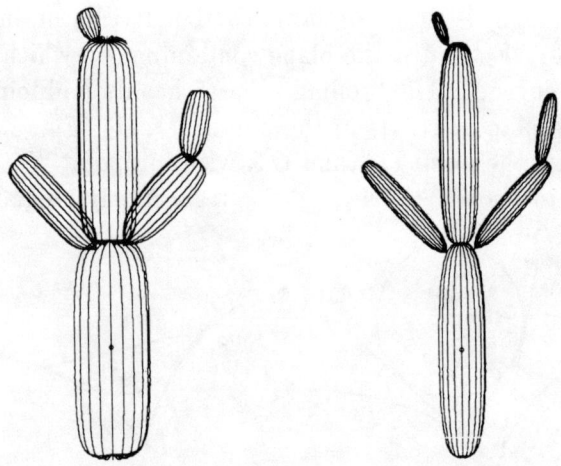

FIG. 3.34. *Modeling of a cactus with different* μ. *The larger* μ *leads to the shape of the cactus after the winter time; the smaller* μ *leads to the shape of the cactus after a hot summer.*

3.4. Modeling with Functional Spline Surfaces

We want to end our research work applying the developed methods to real objects: we will try to simulate the shape of a cactus after the winter period and after a hot summer. We have only to change the factor μ and then we get the representations of a cactus described in Fig. 3.34. Planes were used as bases and transversal surfaces that circumscribe the cactus.

References

[1] M. P. Do Carmo, *Differential Geometry of Curves and Surfaces*, Prentice-Hall, Englewood Cliffs, NJ, 1976.

[2] S. Cohen, *Beitrag zur steuerbaren Interpolation von Kurven und Flächen*, dissertation, TU Dresden, 1982.

[3] J. Favard, *Cours de Géométrie différentielle locale*, Gauthier, Paris, 1957.

[4] Y. Y. Feng, E. Hartmann, and J. Hoschek, *On convexity of functional splines and its application*, in preparation.

[5] Th. Garrity and J. Warren, *Geometric continuity*, Comput. Aided Geom. Des., 8 (1991), pp. 51–66.

[6] J. A. Gregory and J. M. Hahn, *Geometric continuity and convex combination patches*, Comput. Aided Geom. Des., 4 (1987), pp. 79–89.

[7] J. M. Hahn, *Filling polygonial holes with rectangular patches*, in Theory and Practice of Geometric Modeling, W. Strasser and H.-R. Seidel, eds., Springer-Verlag, New York-Berlin, 1990.

[8] E. Hartmann, *Computerunterstützte Darstellende Geometrie*, Teubner, Stuttgart, 1988.

[9] ——, *Blending of implicit surfaces with functional splines*, Comput. Aided Des., 22 (1990), pp. 500–506.

[10] E. Hartmann and J. Li, *Smoothing of corners with functional splines*, preprint, Fachbereich Mathematik, Technische Hochschule Darmstadt, 1989.

[11] Ch. Hoffmann and J. Hopcroft, *Automatic surface generation in computer-aided design*, Visual Computer 1 (1985), pp. 92–100.

[12] ——, *The potential method for blending surfaces and corners*, in Geometric Modelling: Algorithms and New Trends, G. E. Farin, ed., Society for Industrial and Applied Mathematics, Philadelphia, 1987, pp. 347–366.

[13] F. Hohenberg, *Konstruktive Geometrie in der Technik*, Springer-Verlag, New York-Berlin, 1966.

[14] I. Holmström, *Piecewise quadric blending of implicitly defined surfaces*, Comput. Aided Geom. Des., 4 (1987), pp. 171–190.

[15] J. Hoschek, E. Hartmann, J. Li, and Y. Y. Feng, G^{n-1}-*functional splines for interpolation and approximation of surfaces and solids*, International Series in Numerical Mathematics, 94 (1990), pp. 141–154.

[16] J. Hoschek and D. Lasser, *Grundlagen der geometrischen Datenverarbeitung*, Teubner, Stuttgart, 1989.

[17] J. Hoschek and E. Hartmann, G^{n-1} *functional splines for modeling*, in Geometric Modeling Methods and Applications, H. Hagen and Roller, eds., Springer-Verlag, New York-Berlin, 1990, pp. 185–212.

[18] J. Li, J. Hoschek, and E. Hartmann, G^{n-1} *functional splines for interpolation and approximation of curves and surfaces and solids*, Comput. Aided Geom. Des., 7 (1990), pp. 209–220.

[19] R. A. Liming, *Practical Analytical Geometry with Applications to Aircraft*, Macmillan, New York, 1944.

[20] J. Pegna and F.-E. Wolter, *Geometrical criteria to guarantee second order smoothness of blend surfaces*, Trans. ASME, J. Mech. Des., 114 (1992).

[21] V. Pratt, *Direct least-squares fitting of algebraic surfaces*, Computer Graphics, 21 (1987), pp. 145–151.

[22] A. Ricci, *A constructive geometry for computer graphics*, Computer Journal, 16 (1973), pp. 157–160.

[23] A. W. Roberts and D. E. Varberg, *Convex Functions*, Academic Press, New York, 1973.

[24] T. D. De Rose and Ch. T. Loop, *S-patches: A class of representations for multi-sided surface patches*, Tech. Report 88-05-02, University of Washington, 1988.

[25] M. A. Sabin, *Non-rectangular patches suitable for inclusion in a B-Spline surface*, in Proceeding Eurographic, H. Hagen, ed., North-Holland, 1983, pp. 57–69.

[26] G. Scheffers. *Anwendungen der Differential- und Integralrechnung auf Geometrie*, Vol. I, Springer-Verlag, New York, Berlin, 1922.

[27] ——, *Anwendungen der Differential- und Integralrechnung auf Geometrie*, Vol. II, 3rd ed., Springer-Verlag, New York, Berlin, 1922.

[28] M. Schmidt, *Modellierung und Approximation von Kurven und Flächen in impliziter Darstellung*, dissertation, Manuskript, Fachbereich Mathematik Darmstadt, 1992.

[29] B. Spain, *Analytical Quadrics*, Pergamon Press, Elmsford, NY, 1960.

[30] O. Veblen and J. W. Young, *Projective Geometry* I, Ginn, Boston, 1910.

[31] J. R. Woodward, *Blends in geometric modelling*, in The Mathematics of Surfaces II, R. R. Martin, ed., Clarendon Press, Oxford, 1987, pp. 255–297.

Variational Surface Design

A Local Twist Estimator

Gerald Farin and Hans Hagen

4.1. Introduction

The purpose of this paper is to present a new twist estimation for smooth surface design based upon a calculus of variation approach. Surfaces designed in a computer graphic environment have many applications, including the design of cars, airplanes, shipbodies, and modeling robots. The choice of the surface form depends upon the application.

The twist vector of a surface $X(u,v)$ is the mixed partial $X_{uv}(u,v)$. As an example, we give the twist vector of a Bézier patch, given by

$$X(u,v) = \sum_{i=0}^{m}\sum_{j=0}^{n} b_{i,j} B_i^m(u) B_j^n(v), \quad 0 \le u,v \le 1,$$

where the B_i^m are Bernstein polynomials $B_i^m(u) = \binom{m}{i} u^i (1-u)^{m-i}$. The twist at the point $X(0,0)$ is given by

$$X_{u,v}(0,0) = \frac{\partial^2}{\partial u \partial v} X(u,v)|_{u=v=0} = (m-1)(n-1)(b_{1,1} - b_{1,0} - b_{0,1} + b_{0,0}).$$

Geometrically, the twist vector denotes the deviation of the hyperbolic paraboloid formed by $b_{0,0}, b_{0,1}, b_{1,0}, b_{1,1}$ from the parallelogram formed by $b_{0,0}, b_{0,1}, b_{1,0}$. It measures how "curved" the surface is at $X(0,0)$.

If $b_{1,1}$ is in the tangent plane at $b_{0,0}$, we may obtain a nonzero twist vector, yet the surface is not "curved" at $b_{0,0}$. Thus, the "curviness" is determined by the normal component of the twist, not the twist itself. This normal component, called h_{12}, is given by $h_{12} = \langle X_{uv}(u,v), N(u,v) \rangle$, where N is the normal vector at $X(u,v)$. It is discussed in the next section.

Several surface generation methods (Coons patches or bicubic patches) require twist vectors as input data. Since this information is typically not available, one has to estimate the twist of the surface at the patch corners from the four boundary curves of the patch. Many methods exist for this purpose; the reader is referred to [1], [2], [4], or [6] for an overview.

We finish this section with a warning: the simplest twist estimation method, setting the corner twists to zero, is in general also the worst one. It produces surfaces that have flat spots at the patch corners.

4.2. Variational Twist Estimation

Before we can describe this method, some differential geometry of surfaces is needed. For more details, see [3] or W. Böhm in [4]. Let $X(u, v)$ be a sufficiently often differentiable surface. Then, two important geometric quantities are the first and second fundamental matrices G and H:

$$G = \begin{pmatrix} \langle X_u, X_u \rangle & \langle X_u, X_v \rangle \\ \langle X_v, X_u \rangle & \langle X_v, X_v \rangle \end{pmatrix} = \begin{pmatrix} g_{11} & g_{12} \\ g_{21} & g_{22} \end{pmatrix},$$

$$H = \begin{pmatrix} \langle N, X_{uu} \rangle & \langle N, X_{uv} \rangle \\ \langle N, X_{vu} \rangle & \langle N, X_{vv} \rangle \end{pmatrix} = \begin{pmatrix} h_{11} & h_{12} \\ h_{21} & h_{22} \end{pmatrix}.$$

Note that the twist vector appears in the second fundamental matrix.

An important geometric measure for surfaces is the *Gaussian curvature K*:

$$K = \frac{det\ H}{det\ G} = \frac{h}{g}.$$

It describes the local shape of a surface: points with $K > 0, K = 0, K < 0$ are called elliptic, parabolic, or hyperbolic points, respectively. For more details, see [5] or [7].

Another curvature measure is given by the *mean curvature M*:

$$M = \frac{h_{11}g_{22} - 2h_{12}g_{12} + h_{22}g_{11}}{2g}.$$

The mean curvature measures the deviation of a surface from a minimal surface: those are surfaces with mean curvatures equal to zero everywhere. Minimal surfaces correspond to surfaces formed by a soap bubble between boundary curves.

At a given point on a surface, the surface assumes a minimal and maximal normal section curvature, called k_{min} and k_{max}. These extreme curvatures are given by

$$k_{max} = K + \sqrt{M^2 - K},$$
$$k_{min} = K - \sqrt{M^2 - K}.$$

The functional

$$F = \int_s (k_{min}^2 + k_{max}^2)\ ds$$

is a standard fairness criterion for surfaces in engineering [10]. It is used because it measures the strain energy of flexure and torsion in a thin rectangular elastic plate with small deflection. Hagen and Schulze [9] used this functional for a variational formulation of the twist estimation problem, with the assumption of orthogonal patch boundary curves. One can drop this restriction and obtain an estimate for the twist normal component h_{12} as follows.

One verifies

$$F = \int_A \frac{(g_{11}h_{22} - 2g_{12}h_{12} + g_{22}h_{11})^2 - 2gh}{g^2} \sqrt{g}\, du\, dv$$

$$= \int_A [h_{12}^2(2g + 4g_{12}^2) - 4h_{12}g_{12}(g_{11}h_{22} + g_{22}h_{11}),$$

$$+ (g_{11}h_{22} + g_{22}h_{11})^2 - 2gh_{11}h_{22}]g^{-\frac{3}{2}}\, du\, dv$$

where A is the domain surface.

Since we have a functional of the form $F : \int\int f(u, v, h_{12}(u, v))\, du\, dv$, the Euler equation

$$\frac{\partial f}{\partial g_{12}} = 2h_{12}(2g + 4g_{12}^2) - 4g_{12}(g_{11}h_{22} + g_{22}h_{11}) = 0$$

gives a necessary condition for the energy minimum:

$$(4.1) \qquad h_{12} = \frac{g_{12}(g_{11}h_{22} + g_{22}h_{11})}{g + 2g_{12}^2} = 0.$$

We note an interesting relation between this formula and the one for the mean curvature M. It can be written as

$$M = \frac{g_{11}h_{22} + g_{22}h_{11} - 2h_{12}g_{12}}{2g}.$$

Thus, (4.1) is equivalent to

$$h_{12} = 2g_{12}M.$$

We summarize the above in the following theorem.

THEOREM 4.2.1. *A surface that satisfies* (4.1) *at every point is smooth in the sense that*

$$\int_s (k_{\min}^2 + k_{\max}^2)\, ds \quad \to \quad \min.$$

Based on this result, it seems reasonable to base the corner twists X_{uv} on (4.1). We note that (4.1) only gives the normal component of X_{uv}, so we still have to compute its two tangential components.

4.3. Remarks

1. If the patch boundary curves meet at right angles ($g_{12} = 0$ at all patch corners), the theorem yields $h_{12} = 0$. Thus, the boundary curves are lines of curvature near the patch corners. This is a very important property that is not shared by other twist estimators.

2. If the surface contains two asymptotic directions (two families of straight lines, for example), we may locally have asymptotic isoparametric lines with

$h_{11} = h_{22} = 0$. Such a situation is characterized by locally hyperbolic curvature, and our method will produce $h_{12} = 0$. This may lead to flat spots at the patch corners.

3. Our twist scheme is the most local of all known methods, it only considers data at a patch corner to generate the twist there. Thus, the change in data at one data point only affects four points with our method, as opposed to 16 with other methods.

Considering all the advantages and (minor) disadvantages of our variational twist estimation, it is rather straightforward to extend the stiffness degree concept for curves (see [9]) to surfaces.

We assume that the reader is familiar with the Adini-twist concept (see [2]). Performing the blended optimization of the normal components of the energy and of the Adini-twist, and using the (less important) tangent components of the Adini-twist vector $Ad(u, v)$ we get:

$$(4.2) \qquad X_{uv} = (\alpha \langle Ad, N \rangle + \beta h_{12})N + \langle AD, X_u \rangle X_u + \langle Ad, X_v \rangle X_v,$$

$\alpha(u, v), \beta(u, v) \geq 0; \alpha + \beta = 1$. This is the Gauß-frame representation of an (in our meaning) optimal twist vector.

4.4. An Example

We constructed a turbine blade using this method. We also constructed a surface using zero twists. We compared both surfaces by performing the reflection line analysis method (for more details see [2]).

Our twist estimation method clearly gives a smoother surface than the zero twist method, as is illustrated in Figs. 4.1–4.3.

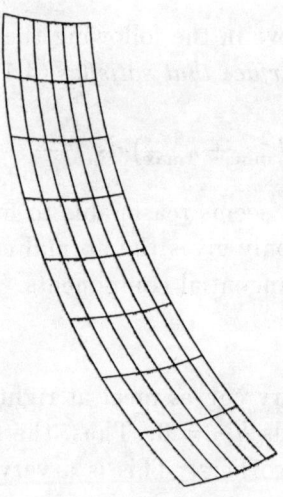

FIG. 4.1. *Turbine blade.*

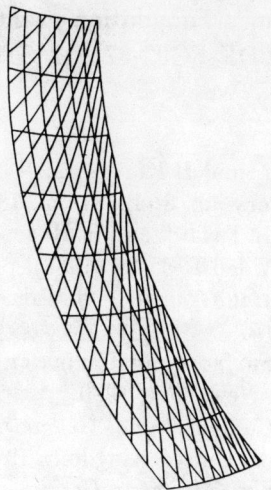

FIG. 4.2. *Reflection-line analaysis of the turbine blade with zero twist input.*

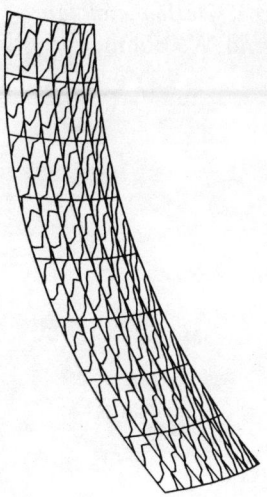

FIG. 4.3. *Reflection-line analysis of the turbine blade with our smooth twist input.*

Acknowledgments

Both authors were supported by DFG grant BR 118/130-1. Farin was also supported by National Science Foundation grant DCR-8502858 and by U.S. Department of Energy grant DE-FG02-87ER25041 to Arizona State University.

References

[1] R. E. Barnhill, J. Brown, and I. Klucewicz, *A new twist in CAGD*, Computer Graphics and Image Processing, 8 (1978), pp. 78–91.

[2] R. E. Barnhill, G. Farin, L. Fayard, and H. Hagen, *Twists, curvatures and surface interrogation*, Comput. Aided Des., 20 (1988), pp. 341–346.

[3] M. do Carmo, *Differential Geometry of Curves and Surfaces*, Prentice-Hall, Englewood Cliffs, NJ, 1976.

[4] G. Farin, *Curves and Surfaces for Computer Aided Geometric Design*, 2nd edition, Academic Press, New York, 1990.

[5] I. Faux and M. Pratt, *Computational Geometry for Design and Manufacture*, Ellis Horwood, Chichester, United Kingdom, 1979.

[6] H. Hagen, *Computer Aided Geometric Design—Methods and Applications*, Proceedings of the Conference on Engineering Graphics and Descriptive Geometry, Wien 1988.

[7] ——, *Geometric surface patches without twist constraints*, Comput. Aided Geom. Des., 3 (1986), pp. 179–184.

[8] H. Hagen and G. Schulze, *Automatic smoothing with geometric surface patches*, Comput. Aided Geom. Des., 4 (1987), pp. 231–236.

[9] ——, *Variational principles in curve and surface design*, in Geometric Modelling—Methods and Applications, D. Roller and G. Farin, eds., Springer-Verlag, New York–Berlin, 1990.

[10] H. Nowacki and D. Reese, *Design and fairing of ship surfaces*, in Surfaces in CAGD, R. E. Barnhill and W. Böhm, eds., North-Holland, Amsterdam, 1983, pp. 121–134.

Variational Design of Smooth B-Spline
Surfaces

Hans Hagen and Paolo Santarelli

5.1. Introduction

Computer aided geometric design (CAGD) has emerged from the need for free-form surfaces in CAD/CAM technologies; it has become a major research topic in computer science with direct applications for all engineering sciences in the last few years. A central problem of geometric modeling is the construction of "technically smooth" surface representations (i.e., the data should be directly usable in NC processing).

In many applications only point data are available, and it is difficult to estimate additional information such as tangents or curvatures. Furthermore, measurement of inaccuracies which occur, for example, when digitizing an original model with worn out parts, must be considered.

The purpose of this paper is to present an algorithm appropriate for these situations. The construction algorithm combines a weighted least-square fit with an automatic smoothing process, based upon a calculus of variation approach and a stiffness degree concept.

5.2. Bézier and B-Spline Surfaces

The curves and surfaces now known as Bézier curves and surfaces were independently developed by P. de Casteljau and P. Bézier. The underlying mathematical theory based on the concept of Bernstein's polynomials was first introduced by R. Forrest (see [2]). The fundamental idea of this approach is to evaluate and manipulate the curves and surfaces by a (small) number of control points. We first consider Bézier curves as segmented curves. The segments $X_l(u); l = 0, \cdots, k$ of a Bézier curve of degree m over the parameter interval $u_l \leq u \leq u_{l+1}$ are

$$(5.1) \qquad X_l(u) := \sum_{i=0}^{m} b_{lm+i} B_i^m \left(\frac{u - u_l}{u_{l+1} - u_l} \right).$$

The Bernstein polynomials $B_i^m(t) := \binom{m}{i}(1-t)^{m-i} t^i$, $0 \leq t \leq 1$ are used as blending functions.

Bernstein polynomials are special degenerated B-splines (see [1]). If we use B-splines as blending functions instead of Bernstein polynomials, we can generalize the whole concept to so-called B-spline curves and surfaces (see [3]). B-spline curves are similar to Bézier curves in that a set of blending functions combine the effect of $n + 1$ control points

$$(5.2) \qquad Y(u) := \sum_{j=0}^{n} d_j N_j^M(u).$$

The most important difference is the local support property of the B-spline blending functions $N_j^k(u)$. Both curve types have the convex hull and variation diminishing property (for more details see [1]).

A Bézier surface is a segmented surface. The segments $X_{pq}(u,v); p = 0, \cdots, k; q = 0, \cdots, r$ of a Bézier surface of degree m, n over the rectangular parameter domain $u_p \leq u \leq u_{p+1}; v_q \leq v \leq v_{q+1}$ are

$$(5.3) \quad X_{pq}(u,v) := \sum_{i=0}^{m} \sum_{j=0}^{n} b_{p \cdot m+i/q \cdot n+j} \cdot B_i^m \left(\frac{u - u_p}{u_{p+1} - u_p} \right) B_j^n \left(\frac{v - v_q}{v_{q+1} - v_q} \right).$$

Instead of a control polygon a Bézier surface (segment) has a control polyhedron.

The definition of a B-spline surface over a rectangular parameter domain directly follows the same pattern,

$$(5.4) \qquad Y(u,v) := \sum_{i=0}^{m} \sum_{j=0}^{n} d_{ij} N_i^M(u) N_j^N(v).$$

The so-called tensor-product surfaces can be easily generated by applying the de Casteljau algorithm or the de Boor algorithm twice.

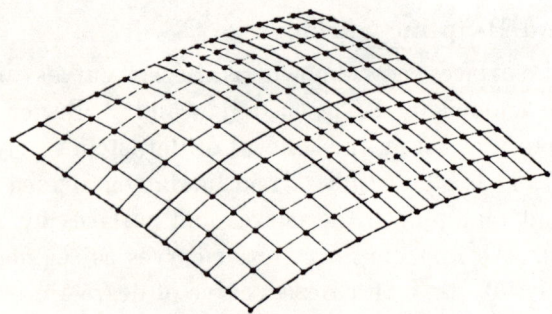

FIG. 5.1.   *B-spline surface.*

5.3. Variational Design of B-Spline Surfaces

In this section we leave the classical approach of (1) constructing a smooth net of curves and (2) adding the surface patches smoothly into the net and present a direct method to construct a technically smooth B-spline surface, which uses only point data and refrains from determining a net.

The construction algorithm combines a weighted least square approximation with automatic surface smoothing. The smoothing criterion is the approximate minimization of the curvature variation. This technique presented here aims at constructing tangent-plane continuous B-spline surfaces. The following mathematical model serves as variation principle:

$$
(5.5) \quad
\begin{aligned}
&(1 - ws)\left\{ \sum_{k=1}^{n_p} w_{pk}\left[X(u_k, v_k) - P_k \right]^2 \right\} \\
&+ ws\left\{ \sum_{i=1}^{n} \sum_{j=1}^{m} w_{3u} \int_{v_j}^{v_{j+1}} \int_{u_i}^{u_{i+1}} w_{3u_{ij}} \left\| \frac{\partial^3 X(u,v)}{\partial u^3} \right\|^2 du\, dv \right. \\
&\left. + w_{3v} \int_{v_j}^{v_{j+1}} \int_{u_i}^{u_{i+1}} w_{3v_{ij}} \left\| \frac{\partial^3 X(u,v)}{\partial v^3} \right\|^2 du\, dv \right\} \quad \rightarrow \quad \min.
\end{aligned}
$$

$X(u, v)$ is the representation of the surface; $(u, v) \in [u_1, u_{n+1}] \times [v_1, v_{m+1}]$ is the parameter value; and n, m are the number of segments in a u and v direction. P_k are the points to be approximated and n_p is the number of these points. The weight coefficients ws, w_{3u}, w_{3v}, $w_3 u_{ij}$, $w_3 v_{ij}$ are valid in the interval $[0, 1]$ and fulfill the constraints $\sum_{i=1}^{m} \sum_{j=1}^{n} w_{3u_{ij}} = 1$ and $\sum_{i=1}^{m} \sum_{j=1}^{n} w_{3v_{ij}} = 1$.

We apply this variation principle to biquintic B-spline surfaces,

$$
(5.6) \quad X(u, v) = \sum_{i=1}^{4n+2} \sum_{j=1}^{4m+2} d_{ij} N_i^5(u) N_j^5(v),
$$

with the knot-vectors

$$
\{ k_1^u, \cdots, k_{4n+8}^u \}
$$

$$
:= \{ \underbrace{u_1, \cdots, u_1}_{6*}, \underbrace{u_2, u_2, u_2, u_2}_{4*}, \cdots, \underbrace{u_n, u_n, u_n, u_n}_{4*}, \underbrace{u_{n+1}, \cdots, u_{n+1}}_{6*} \}
$$

and

$$
\{ k_1^v, \cdots, k_{4m+8}^v \}
$$

$$
:= \{ \underbrace{v_1, \cdots, v_1}_{6*}, \underbrace{v_2, v_2, v_2, v_2}_{4*}, \cdots, \underbrace{v_m, v_m, v_m, v_m}_{4*}, \underbrace{v_{m+1}, \cdots, v_{m+1}}_{6*} \}.
$$

This set of knot vectors guarantees the C^1-continuity of the surface.

We can now use the control points d_{ij}; $i \in \{1, \cdots, 4n+2\}$, $j \in \{1, \cdots, 4m+ 2\}$ as parameters for the calculus of variation approach.

Applying the variation principle (5.5) is a three-step process.

Step 1. Least square fitting.

$$
(5.7) \quad SL := \sum_{k=1}^{n_p} w_{pk}[F(u_k, v_k) - P_k]^2 \quad \rightarrow \quad \min,
$$

or in B-spline representation

$$(5.8) \qquad \sum_{k=1}^{n_p} w_{pk} \left[\sum_{l=1}^{4n+2} \sum_{r=1}^{4m+2} d_{lr} N_l^5(u_k) N_r^5(v_k) - P_k \right]^2 \rightarrow \quad \min .$$

The necessary conditions $\frac{\partial SL}{\partial d_{ij}} = 0$ lead to a linear system of equations:

$$(5.9) \qquad \sum_{l=1}^{4n+2} \sum_{r=1}^{4m+2} \left\{ 2 \sum_{k=1}^{n_p} w_{pk} \cdot N_l^5(u_k) N_r^5(v_k) N_i^5(u_k) N_j^5(v_k) \right\} d_{lr}$$

$$= 2 \sum_{k=1}^{n_p} w_{pk} \cdot P_k N_i^5(u_k) N_j^5(v_k).$$

The unique solution of this system is the best point fitting in the least square sense of (5.7).

Step 2. Automated smoothing process. As a fairness criterion we use

$$(5.10) \qquad \sum_{i=1}^{n} \sum_{j=1}^{m} \int_{v_j}^{v_j+1} \int_{u_i}^{u_i+1} w3u \cdot w3u_{ij} \left\| \frac{\partial^3 X(u,v)}{\partial u^3} \right\|^2$$

$$+ w3v \cdot w3v_{ij} \left\| \frac{\partial^3 X(u,v)}{\partial v^3} \right\|^2 du\, dv \rightarrow \min .$$

A calculus of variation approach leads for the first term to a linear system of equations,

$$\frac{\partial I30(v)}{\partial d_{ij}} = (F4U_i \cdot C1U_0) d_i(v) \cdot N_j(v)$$

$$+ (F5U_i \cdot C1U_1 + F4U_i \cdot C2U_0) d_2(v) \cdot N_j(v)$$
$$+ (F6U_i \cdot C1U_2 + F5V_i \cdot C2U_1 + F4U_i \cdot C3U_0) d_3(v) \cdot N_j(v)$$
$$+ (F7U_i \cdot C1U_3 + F6U_i \cdot C2U_2 + F5U_i \cdot C3U_1) d_4(v) \cdot N_j(v)$$
$$+ (F7U_i \cdot C2U_3 + F6U_i \cdot C3U_2) d_5(v) \cdot N_j(v)$$
$$+ (F7U_i \cdot C3U_3) d_6(v) \cdot N_j(v), \qquad i = 1,2.$$

$$\frac{\partial I30(v)}{\partial d_{ij}} = (F0U_i \cdot C1U_{4l-4}) d_{4l-3}(v) \cdot N_j(v) + (F1U_i \cdot C1U_{4l-3}$$

$$+ F0U_i \cdot C2U_{4l-4}) d_{4l-2}(v) \cdot N_j(v) + (F2U_i \cdot C1U_{4l-2}$$

$$+ F1U_i \cdot C2U_{4l-3} + F0U_i \cdot C3U_{4l-4}) d_{4l-1}(v) \cdot N_j(v)$$

$$+ (F3U_i \cdot C1U_{4l-1} + F2U_i \cdot C2U_{4l-2} + F1U_i \cdot C3U_{4l-3}) d_{4l}(v)$$

$$\cdot N_j(v) + (F3U_i \cdot C2U_{4l-1} + F2U_i \cdot C3U_{4l-2}) d_{4l+1}(v) \cdot N_j(v)$$

$$+ (F3U_i \cdot C3U_{4l-1}) d_{4l+2}(v) \cdot N_j(v),$$

$$i = 4l - 1, 4l, \quad l \in \{1, \cdots, n\}.$$

$$\frac{\partial I30(v)}{\partial d_{ij}} = (F0U_i \cdot C1U_{4l-4})d_{4l-3}(v) \cdot N_j(v) + (F1U_i \cdot C1U_{4l-3}$$

$$+ F0U_i \cdot C2U_{4l-4})d_{4l-2}(v) \cdot N_j(v) + (F2U_i \cdot C1U_{4l-2}$$

$$+ F1U_i \cdot C2U_{4l-3} + F0U_i \cdot C3U_{4l-4})d_{4l-1}(v) \cdot N_j(v)$$

$$+ (F3U_i \cdot C1U_{4l-1} + F2U_i \cdot C2U_{4l-2} + F1U_i \cdot C3U_{4l-3})d_{4l}(v)$$

$$\cdot N_j(v) + (F4U_i \cdot C1U_{4l} + F3U_i \cdot C2U_{4l-1}$$

$$+ F2U_i \cdot C3U_{4l-2})d_{4l+1}(v) \cdot N_j(v) + (F5U_i \cdot C1U_{4l+1}$$

$$+ F4U_i \cdot C2U_{4l} + F3U_i \cdot C3U_{4l-1})d_{4l+2}(v) \cdot N_j(v)$$

$$+ (F6U_i \cdot C1U_{4l+2} + F5U_i \cdot C2U_{4l+1} + F4U_i \cdot C3U_{4l})d_{4l+3}(v)$$

$$\cdot N_j(v) + (F7U_i \cdot C1U_{4l+3} + F6U_i \cdot C2U_{4l+2}$$

$$+ F5U_i \cdot C3U_{4l+1})d_{4l+4}(v) \cdot N_j(v) + (F7U_i \cdot C2U_{4l+3}$$

$$+ F6U_i \cdot C3U_{4l+2})d_{4l+5}(v) \cdot N_j(v) + (F7U_i \cdot C3U_{4l+3})d_{4l+6}(v)$$

$$\cdot N_j(v), \qquad i = 4l+1, 4l+2, \quad l \in \{1, \cdots, n-1\}.$$

$$\frac{\partial I30(v)}{\partial d_{ij}} = (F0U_i \cdot C1U_{4n-4})d_{4n-3}(v) \cdot N_j(v) + (F1U_i \cdot C1U_{4n-3}$$

$$+ F0U_i \cdot C2U_{4n-4})d_{4n-2}(v) \cdot N_j(v) + (F2U_i \cdot C1U_{4n-2}$$

$$+ F1U_i \cdot C2U_{4n-3} + F0U_i \cdot C3U_{4n-4})d_{4n-1}(v) \cdot N_j(v)$$

$$+ (F3U_i \cdot C1U_{4n-1} + F2U_i \cdot C2U_{4n-2} + F1U_i \cdot C3U_{4n-3})d_{4n}(v)$$

$$\cdot N_j(v) + (F3U_i \cdot C2U_{4n-1} + F2U_i \cdot C3U_{4n-2})d_{4n+1}(v) \cdot N_j(v)$$

$$+ (F3U_i \cdot C3U_{4n-1})d_{4n+2}(v) \cdot N_j(v),$$

$$i = 4n+1, 4n+2, \quad j \in \{1, \cdots, 4n+2\}.$$

$$C3U_j = \frac{4}{K^n_{j+7} - K^n_{j+3}} \cdot \frac{5}{K^n_{j+8} - K^n_{j+3}},$$

$$C2U_j = -\frac{4}{K_{j+7} - K_{j+3}} \left(\frac{5}{K^n_{j+8} - K^n_{j+3}} + \frac{5}{K^n_{j+7} - K^n_{j+2}} \right),$$

$$C3U_j = \frac{4}{K^n_{j+7} - K^n_{j+3}} \cdot \frac{5}{K^n_{j+7} - K^n_{j+2}},$$

$$F4U_1 := \tfrac{9}{5} \cdot \tfrac{w3ug}{\Delta u_1} \cdot (2 \cdot C1U_0),$$

$$F5U_1 := \tfrac{9}{5} \cdot \tfrac{w3ug}{\Delta u_1} \cdot (-1 \cdot C1U_0),$$

$$F6U_1 := \tfrac{9}{5} \cdot \tfrac{w3ug}{\Delta u_1} \cdot \left(-\tfrac{2}{3} \cdot C1U_0\right),$$

$$F7U_1 := \tfrac{9}{5} \cdot \tfrac{w3ug}{\Delta u_1} \cdot \left(-\tfrac{1}{3} \cdot C1U_0\right),$$

$$F4U_2 := \tfrac{9}{5} \cdot \tfrac{w3ug}{\Delta u_1} \cdot (2 \cdot C2U_0 - C1U_1),$$

$$F5U_2 := \tfrac{9}{5} \cdot \tfrac{w3ug}{\Delta u_1} \cdot \left(-C2U_0 + \tfrac{4}{3} \cdot C1U_1\right),$$

$$F6U_2 := \tfrac{9}{5} \cdot \tfrac{w3ug}{\Delta u_1} \cdot \left(-\tfrac{2}{3} \cdot C2U_0 + \tfrac{1}{3} \cdot C1U_1\right),$$

$$F7U_2 := \tfrac{9}{5} \cdot \tfrac{w3ug}{\Delta u_1} \cdot \left(-\tfrac{1}{3} \cdot C2U_0 - \tfrac{2}{3} \cdot C1U_1\right),$$

$$F0U_{4l-1} := \tfrac{9}{5} \cdot \tfrac{w3ug}{\Delta u_l} \cdot \left(-2 \cdot C3U_{4(l-1)} - C2U_{4(l-1)+2} - \tfrac{2}{3} \cdot C1U_{1(l-1)+2} \right),$$

$$F1U_{4l-1} := \tfrac{9}{5} \cdot \tfrac{w3ug}{\Delta u_l} \cdot \left(-C3U_{4(l-1)} + \tfrac{4}{3} \cdot C24_{4(l-1)+1} + \tfrac{1}{3} \cdot C1U_{4(l-1)+2} \right),$$

$$F2U_{4l-1} := \tfrac{9}{5} \cdot \tfrac{w3ug}{\Delta u_l} \cdot \left(-\tfrac{2}{3} \cdot C3U_{4(l-1)} + \tfrac{1}{3} \cdot C2U_{4(l+1)+1} + \tfrac{4}{3} \cdot C1U_{4(l+1)+2} \right),$$

$$F3U_{4l-1} := \tfrac{9}{5} \cdot \tfrac{w3ug}{\Delta u_l} \cdot \left(-\tfrac{1}{3} \cdot C3U_{4(l-1)+2} - \tfrac{2}{3} \cdot C2U_{4(l-1)+1} - C1U_{4(l-1)+2} \right),$$
$$\{l \in 1, \cdots, n\}.$$

$$F0U_{4l} := \tfrac{9}{5} \cdot \tfrac{w3ug}{\Delta u_l} \cdot \left(-C3U_{4(l-1)+1} - \tfrac{2}{3} \cdot C2U_{4(l-1)+2} - \tfrac{1}{3} \cdot C1U_{4(l-1)+3} \right),$$

$$F1U_{41} := \tfrac{9}{5} \cdot \tfrac{w3ug}{\Delta u_l} \cdot \left(-\tfrac{4}{3} \cdot C3U_{4(l-1)+1} + \tfrac{1}{3} \cdot C24_{4(l-1)+2} - \tfrac{2}{3} \cdot C1U_{4(l-1)+3} \right),$$

$$F2U_{4l} := \tfrac{9}{5} \cdot \tfrac{w3ug}{\Delta u_l} \cdot \left(-\tfrac{1}{3} \cdot C3U_{4(l-1)+1} + \tfrac{4}{3} \cdot C2U_{4(l+1)+2} - C1U_{4(l-1)+3} \right),$$

$$F3U_{41} := \tfrac{9}{5} \cdot \tfrac{w3ug}{\Delta u_l} \cdot \left(-\tfrac{2}{3} \cdot C3U_{4(l-1)+1} - C2U_{4(l-1)+2} + 2 \cdot C1U_{4(l-1)+3} \right),$$
$$l \in \{1, \cdots, n\}.$$

$$F0U_{4l+1} := \tfrac{9}{5} \cdot \tfrac{w3ug}{\Delta u_l} \cdot \left(-\tfrac{2}{3} \cdot C3U_{4(l-1)+2} - \tfrac{1}{3} \cdot C2U_{4(l-1)+3} \right),$$

$$F1U_{41+1} := \tfrac{9}{5} \cdot \tfrac{w3ug}{\Delta u_l} \cdot \left(\tfrac{1}{3} \cdot C3U_{4(l-1)+2} - \tfrac{2}{3} \cdot C24_{4(l-1)+3} \right),$$

$$F2U_{4l+1} := \tfrac{9}{5} \cdot \tfrac{w3ug}{\Delta u_l} \cdot \left(-\tfrac{4}{3} \cdot C3U_{4(l-1)+2} - C2U_{4(l-1)+3} \right),$$

$$F3U_{41+1} := \tfrac{9}{5} \cdot \tfrac{w3ug}{\Delta u_l} \cdot \left(-C3U_{4(l-1)+2} + 2 \cdot C2U_{4(l-1)+3} \right),$$

$$F4U_{4l+1} := \tfrac{9}{5} \cdot \tfrac{w3ug}{\Delta u_l} \cdot (2 \cdot C1U_{4l}),$$

$$F5U_{4l+1} := \frac{9}{5} \cdot \frac{w3ug}{\Delta u_l} \cdot (-C1U_{4l}),$$

$$F6U_{4l+1} := \frac{9}{5} \cdot \frac{w3ug}{\Delta u_l} \cdot \left(-\frac{2}{3} \cdot C1U_{4l}\right),$$

$$F7U_{4l+1} := \frac{9}{5} \cdot \frac{w3ug}{\Delta u_l} \cdot \left(-\frac{1}{3} \cdot C1U_{4l}\right), \qquad l \in \{1, \cdots, n-1\}.$$

$$F0U_{4l+2} := \frac{9}{5} \cdot \frac{w3ug}{\Delta u_l} \cdot \left(-\frac{1}{3} \cdot C3U_{4(l-1)+3}\right),$$

$$F1U_{4l+2} := \frac{9}{5} \cdot \frac{w3ug}{\Delta u_l} \cdot \left(-\frac{2}{3} \cdot C3U_{4(l-1)+3}\right),$$

$$F2U_{4l+2} := \frac{9}{5} \cdot \frac{w3ug}{\Delta u_l} \cdot \left(-C3U_{4(l-1)+3}\right),$$

$$F3U_{4l+2} := \frac{9}{5} \cdot \frac{w3ug}{\Delta u_l} \cdot \left(2 \cdot C3U_{4(l-1)+3}\right),$$

$$F4U_{4l+2} := \frac{9}{5} \cdot \frac{w3ug}{\Delta u_l} \cdot \left(2 \cdot C2U_{4l} - C1U_{4l+1}\right),$$

$$F5U_{4l+2} := \frac{9}{5} \cdot \frac{w3ug}{\Delta u_l} \cdot \left(-C2U_{4l} + \frac{4}{3} \cdot C1U_{4l+1}\right),$$

$$F6U_{4l+2} := \frac{9}{5} \cdot \frac{w3ug}{\Delta u_l} \cdot \left(\frac{2}{3} \cdot C2U_{4l} + \frac{1}{3} \cdot C1U_{4l+1}\right),$$

$$F7U_{4l+2} := \frac{9}{5} \cdot \frac{w3ug}{\Delta u_l} \cdot \left(\frac{1}{3} \cdot C2U_{4l} - \frac{2}{3} \cdot C1U_{4l+1}\right), \qquad l \in \{1, \cdots, n-1\}.$$

$$F0U_{4n+1} := \frac{9}{5} \cdot \frac{w3ug}{\Delta u_n} \cdot \left(-\frac{2}{3} \cdot C3U_{4(n-1)+2} - \frac{1}{3} \cdot C2U_{4(n-1)+3}\right),$$

$$F1U_{4n+1} := \frac{9}{5} \cdot \frac{w3ug}{\Delta u_n} \cdot \left(\frac{1}{3} \cdot C3U_{4(n-1)+2} - \frac{2}{3} \cdot C2U_{4(n-1)+3}\right),$$

$$F2U_{4n+1} := \frac{9}{5} \cdot \frac{w3ug}{\Delta u_n} \cdot \left(\frac{4}{3} \cdot C3U_{4(n-1)+2} - C2U_{4(n-1)+3}\right),$$

$$F3U_{4n+1} := \frac{9}{5} \cdot \frac{w3ug}{\Delta u_n} \cdot \left(-C3U_{4(n-1)+2} + 2 \cdot C2U_{4(n-1)+3}\right),$$

$$F0U_{4n+2} := \frac{9}{5} \cdot \frac{w3ug}{\Delta u_n} \cdot \left(-\frac{1}{3} \cdot C3U_{4(n-1)+3}\right),$$

$$F1U_{4n+2} := \frac{9}{5} \cdot \frac{w3ug}{\Delta u_n} \cdot \left(-\frac{2}{3} \cdot C3U_{4(n-1)+3}\right),$$

$$F2U_{4n+2} := \frac{9}{5} \cdot \frac{w3ug}{\Delta u_n} \cdot \left(-C3U_{4(n-1)+3}\right),$$

$$F3U_{4n+2} := \frac{9}{5} \cdot \frac{w3ug}{\Delta u_n} \cdot \left(2 \cdot C3U_{4(n-1)+3}\right).$$

These equations have to be intergrated with respect to v. Based on the symmetry of the two components of our smoothing criterion, we get for the second term analogues equations with respect to the v direction.

Step 3. Merging. We now combine the weighted least square fit with the automated smoothing process

$$(1 - ws)A + wsB = 0.$$

A symbolizes (5.9) and *B* symbolizes the equations of Step 2.

5.4. Applications

We use this method to construct reflection surfaces for car headlights (see Figs. 5.2–5.4).

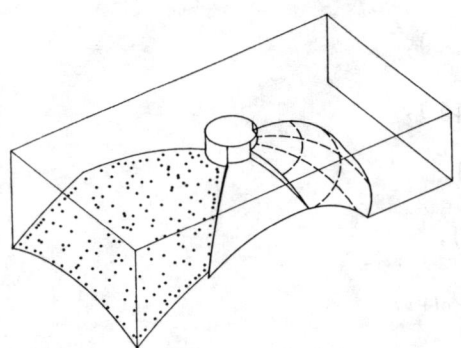

FIG. 5.2.   *Step 1: Digitizing.*

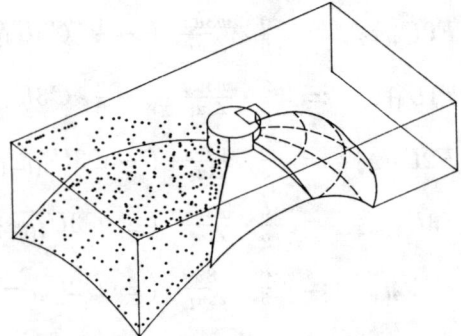

FIG. 5.3.   *Step 2: Parameterization.*

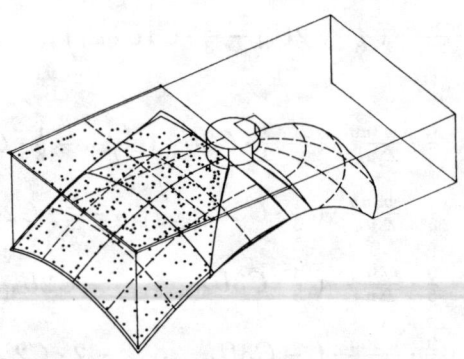

FIG. 5.4.   *Step 3: Variational surface design.*

References

[1] W. Böhm, G. Farin, and J. Kahmann, *A survey of curve and surface methods in CAGD*, Comput. Aided Geom. Des., 1 (1984), pp. 1–60.

[2] R. Forrest, *Interactive interpolation and approximation by Bézier polynomials*, Computer Journal, 15 (1972), pp. 71–79.

[3] W. Gordon and R. Riesenfeld, *B-Spline Curves and Surfaces*, in Computer Aided Geometic Design, R. E. Barnhill and R. Riesenfeld, eds., Academic Press, New York, 1974, pp. 95–126.

Special Applications

Interactive Techniques for Visual Design

Roger K. E. Andersson and Björn E. J. Dahlberg

6.1. Introduction

The process of generating surface descriptions of a quality necessary to serve as a basis for computations as well as manufacturing is a very time consuming task. This is particularly true for surfaces that have to satisfy aesthetic demands, such as styling dependent surfaces within the automotive industry.

Other kinds of surfaces entering into the design process, such as those representing structural sheet metal parts, or the ones used for press tool definitions, also require a substantial effort for their generation, as evidenced by [6]. As a matter of fact, because of its costs, the benefits of the process have been questioned for some purposes.

While the actual time estimations given in [6] quantify the complexity of the problem, it is also of interest to understand the reasons behind this complexity. One important contributor to the difficulty of the task is that most desired surface properties depend in a *nonlinear* fashion on *many variables*. As an example, for a G^1 surface consisting of 12 patches, ordered as 2 adjacent strips with 6 patches in each, even if confined to "transversal" modifications of the control points, the shape depends nonlinearly on 85 variables.

It seems to be a generally accepted rule that a good designer may predict the effect of varying 5 to 6 variables and that an expert may possibly come close to an optimal result, if it depends on 7 to 8 variables. With this in mind, we have to admit that present CAD systems fall short of being an efficient tool for the generation of surfaces other than the simplest ones.

Naturally, the question arises: *How could we reduce the complexity of surface design on the part of the user?*

Whereas the new CAD and design systems are much more user friendly and make the movements of control points easier and their heavily improved graphic capabilities improve the possibility of assessing the effect of such movements, they do not essentially change the complexity of the problem.

Our answer to the question would be to split the task of generation and modifications of surfaces into two parts: *specification* of desired properties and an *automatic fulfillment* of them.

The task of specifying desired properties on a conceptual level is, in general, a creative process that could not be left to any system. The enormously laborious task of fulfillment of such properties, on the other hand, would likely be much more efficiently solved by machine than by man.

For certain requirements that could be set up in advance in a noninteractive manner, like convexity and a fixed degree of smoothness, we have already seen the immense power that can be gained by a system that automatically generates a surface with the required properties (see [1]).

For more refined surface properties, however, a user cannot formulate his or her requirements in a vacuum, but needs a basis formed by a good visualisation, revealing the properties of the current surface. While requirements of this kind must be set up interactively, for their proper formulation a "language for shapes" is needed. The most efficient way to impose such requirements seems to be by modifying the visualization itself.

It is our purpose to discuss some problems that crop up in attempting to create a system of this kind. We will confine the discussion to properties that would naturally be expressed through isophotes or through the vector field of unit normals to the surface.

6.2. The Fundamental Mathematical Results

This section describes some mathematical results related to the visual design of surfaces. Our results serve as a justification for our algorithms for surface modification.

The visual properties of a surface depend on the variation of the normals of the surface. The problem of generating a surface with desired visual properties is therefore a problem of generating a surface whose normals have a prescribed behavior. We will discuss different approaches to this problem. The first is to study the reflexion properties by considering the reflexion lines of the surface. The second is to modify the normals of the surface directly.

Let S be a surface in $\mathbb{R}^3$. For a direction $e \in \mathbb{R}^3 \setminus 0$ and a number $c \in \mathbb{R}$ we define the isophote $\Gamma(S, c, e) = \{p \in S :< n(p), e >= c\}$. We will study the reflexion properties of a surface by considering the isophotes. We let $G(S)$ denote the collection of all unit normals to the surface.

Select a viewing plane and local coordinates (u, v, w) so that the (u, v) coordinates correspond to the viewing plane and the w-direction is the normal to the viewing plane. Let us also assume that the surface S is represented as a graph of a function $f : D \to \mathbb{R}$, where D is a domain in $\mathbb{R}^2$. We remark that the isophotes are the level sets $\{\varphi(u, v) = c\}$ where $\varphi(u, v, e, S) = \varphi(u, v) = (e_3 - e_1 \frac{\partial f}{\partial u} - e_2 \frac{\partial f}{\partial v})(1 + \frac{\partial f}{\partial u}^2 + \frac{\partial f}{\partial u}^2)^{-\frac{1}{2}}$ and e_1, e_2, and e_3 are the components of e in the local coordinates.

In order to build easy-to-use software for the generation of surfaces with prescribed visual properties, it is important to guarantee that there is no need to impose compatibility relations between the input data. Our mathematical results give precise conditions for this.

In this and the next section, we will only be concerned with smooth and strictly convex surfaces. Our first result states that small modifications of the function φ are always possible.

THEOREM 6.2.1. *Let $D \subset \mathbb{R}^2$ be a bounded convex set. Assume that $f_0 : D \to \mathbb{R}$ is strictly convex and belongs to $C^\infty(\bar{D})$. Let S_0 denote the graph of f_0 and assume that e is unit vector such that neither e nor $-e$ belongs to $G(S_0)$. Let $g_0(u,v) = \varphi(u,v,e,S_0)$. If $g \in C^\infty(\bar{D})$ and if g is sufficiently close to g_0 in the topology of $C^\infty(\bar{D})$ then there is a strictly convex function f belonging to $C^\infty(\bar{D})$ such that if S denotes the graph of f then $\varphi(u,v,e,S) = g(u,v)$.*

From this result it is easy to see that small deformations of the isophotes are possible.

THEOREM 6.2.2. *Let S_0 be a strictly convex smooth surface in $\mathbb{R}^3$. Assume that S_0 can be described as a graph over a plane L and let D denote its projection onto L. Suppose e is a unit direction such that neither e nor $-e$ belongs to $G(S_0)$. Let $\Gamma_i = \Gamma(S_0, c_i, e)$, $1 \le i \le n$, be given isophotes and let Γ_i^* be their projections onto L. Suppose γ_i are curves in D. If the curves γ_i are sufficiently close to Γ_i^* in the C^∞ topology, then there is a smooth strictly convex surface M such that the projections of the isophotes $\Gamma(M, c_i, e)$ onto L equal γ_i.*

THEOREM 6.2.3. *Let $D \subset \mathbb{R}^2$ be a bounded convex set. Assume that $f_0 : D \to \mathbb{R}$ is strictly convex and belongs to $C^\infty(\bar{D})$. Let S_0 denote the graph of f_0 and assume that e is a unit vector such that e belongs to $G(S_0)$ and let (u_0, v_0) denote the point whose normal equals e. Set $g_0(u,v) = \varphi(u,v,e,S_0)$. Let (u^*, v^*) be a point in D. Assume $g \in C^\infty(\bar{D})$ and that $g < 1$ in $D \setminus (u^*, v^*)$, $g(u^*, v^*) = 1$ and $g(u,v) \le 1 - C((u-u^*)^2 + (v-v^*)^2)$ for some positive constant C. If (u^*, v^*) is sufficiently close to (u_0, v_0) and if g is sufficiently close to g_0 in the topology of $C^\infty(\bar{D})$ then there is a strictly convex function f belonging to $C^\infty(\bar{D})$ such that if S denotes the graph of f then $\varphi(u,v,e,S) = g(u,v)$.*

The above results are proved by using the Nash-Moser iteration technique, see Hörmander [10].

In the case when isophotes are defined using a normal direction we have, in fact, uniqueness. We remark that in this case, we can choose the direction e as the normal to the viewing plane. If we represent the surface as a graph over this plane, then $\varphi(u,v) = (1 + \| \nabla f \|^2)^{-\frac{1}{2}}$. The uniqueness follows from the following theorem.

THEOREM 6.2.4. *Let D be a bounded convex domain. Suppose f and g are smooth and strictly convex in D. Suppose $\nabla f(u_0, v_0) = 0$. If $g(u_0, v_0) = f(u_0, v_0)$ and if $\| \nabla f \| = \| \nabla g \|$ then $f = g$ in D.*

Because of the compatibility relations involved, it is difficult to prescribe the normals for a surface. A possibility here is to mark the desired changes in the directions of the projections onto the viewing plane of the normals. If we assume that the surface is the graph of a function f over the viewing plane, this corresponds to prescribing $\Lambda f = \arg(\nabla f)$, where $\arg(x)$ denotes the angle between x and the positive u axis.

THEOREM 6.2.5. *Let $D \subset \mathbb{R}^2$ be a bounded convex set. Assume that $f_0 : D \to \mathbb{R}$ is strictly convex, has a nonvanishing gradient and belongs to $C^\infty(\bar{D})$. Let $g_0(u, v) = \Lambda f_0$. If $g \in C^\infty(\bar{D})$ and if g is sufficiently close to g_0 in the topology of $C^\infty(\bar{D})$ then there is a strictly convex function f belonging to $C^\infty(\bar{D})$ such that*

$$\Lambda f = g.$$

6.3. Surface Design Using Reflection Lines

Let S be a smooth and strictly convex surface. We assume that S is defined by a vector $s \in \mathbb{R}^m$. We also assume that for given local coordinates (u, v) the corresponding point in space $P = P(u, v, s)$ is a smooth function of (u, v, s). We remark that if S is defined as a Bézier surface, then s would be all components of all control points.

We again let $G(S)$ denote the collection of all normals of S. For $f \in G(S)$ we denote by $Q(f, S)$ the point $p \in S$ such that the normal to S at p equals f. If M is a surface defined by the vector $s + h \in \mathbb{R}^m$ then

$$(6.1) \qquad\qquad Q(f, M) \approx Q(f, S) + \Sigma \frac{\partial Q}{\partial s_i} h_i.$$

We will next calculate $\frac{\partial Q}{\partial s_i}$. Let $(u(f, S), v(f, S))$ be the local coordinates of the point $Q(f, S)$ and let $n(u, v)$ be the unit normal at the point with local coordinates (u, v). Since $N = n(u(f, S), v(f, S))$ we find by differentiating that

$$(6.2) \qquad\qquad \frac{\partial n}{\partial u} \frac{\partial u(f, S)}{\partial s_i} + \frac{\partial n}{\partial v} \frac{\partial v(f, S)}{\partial s_i} + \frac{\partial n}{\partial s_i} = 0.$$

Since $< n, n >= 1$ it follows by differentiation that $< \frac{\partial n}{\partial s_i}, n >= 0$, so $\frac{\partial n}{\partial s_i}$ is a tangent vector. Since S is strictly convex it follows that $\frac{\partial n}{\partial u}$ and $\frac{\partial n}{\partial v}$ span the tangent plane to S. Therefore relation (6.2) determines $\frac{\partial u(f, S)}{\partial s_i}$ and $\frac{\partial v(f, S)}{\partial s_i}$ uniquely. Since

$$\frac{\partial Q(f, S)}{\partial s_i} = \frac{\partial P}{\partial u} \frac{\partial u(f, S)}{\partial s_i} + \frac{\partial P}{\partial v} \frac{\partial v(f, S)}{\partial s_i} + \frac{\partial P}{\partial s_i}$$

so everything in (6.1) is known.

Now let Γ be an isophote to S and let Π be the projection operator onto the viewing plane. Let γ be a curve in the viewing plane that is close to $\Pi\Gamma$. In order to deform S so that γ becomes the projection of an isophote, we first chose a discrete subset $\{p_j\}$ on $\Pi\Gamma$. We let q_j be the point on γ that is closest to p_j. We next try to solve for h from the linearized equations

$$\Pi \Sigma \frac{\partial Q(f, S)}{\partial s_i} h_i = q_j - p_j.$$

If necessary, other shape constraints are added and the process is then iterated until the desired conditions have been met.

Figure 6.1 displays a surface with a few isophotes. These are modified into a desired pattern and the surface is reshaped by software designed as outlined in this section. Figure 6.2 displays the modified isophotes and the resulting surface.

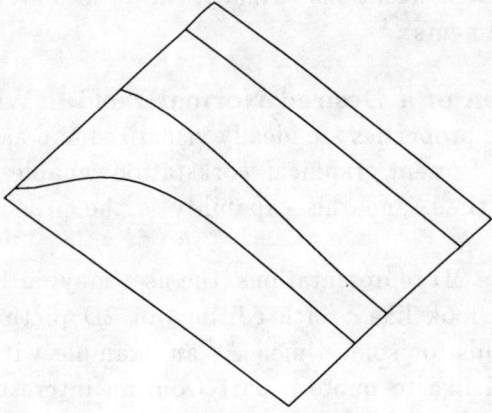

FIG. 6.1. *Isophotes of unmodified surface.*

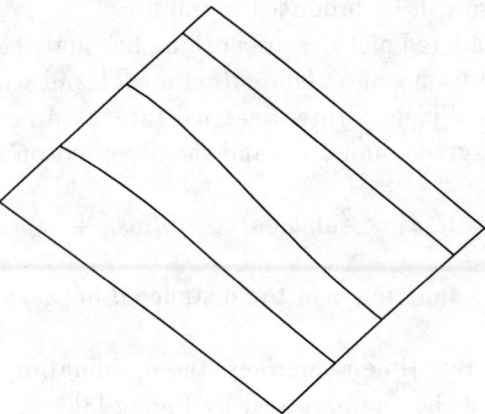

FIG. 6.2. *Isophotes of modified surface.*

6.4. Surface Design Through Normal Fields

An important group of problems arising within computer aided visual design could be summarized as follows: To each point of an existing surface there is associated a unit vector. The surface is to be deformed in such a way that its field of unit normals becomes close to the given unit vector field. In this section, we will study various aspects of this problem.

First we sketch how a desired vector field may originate. We will then introduce a kind of surface description that seems natural for modifications of this sort. After having derived its normal field and approximations to it, we formulate the approximation problem. It is shown that for this problem to admit a precise solution, the given vector field must satisfy a compatibility condition.

The key issue here is that the mathematical results already obtained guarantee that this condition is satisfied for the vector fields that emerge from carefully chosen design situations. Finally, the results are elucidated by a set of numerical experiments.

6.4.1. Generation of a Desired Normal Field.

Within a certain level of precision, surface properties are ideally visualized and assessed through high quality renderings. Present graphical workstations enable such 2D pictures to be formed in real time, and this capability is the principal ingredient in a modern design system.

By changing the 2D representations, the user may fairly well express what the picture should look like. Such editings of 2D pictures have been used within different fields for some time. As an example within the automotive industry, we would like to quote a part from an interesting treatment [14]: "Ford, for example, use the Dubner paint system to create CAD rendered images with 'airbrush effect' colour shading. Such systems have remarkable characteristics—they can assimilate and distort pictures of existing vehicles so that 'restyled' design can be produced in minutes."

Basically, 2D-rendered pictures are nothing but matrices of *image-intensity values*, $I(\lambda)$, formed from some *illumination model*, and with λ the wavelength (see, e.g., [8]). Usually, just three matrices are used, corresponding to the wavelengths for red, green, and blue, and the illumination model is of the form

$$I(\lambda) = f(d) \times (ambient + \ diffuse + \ specular),$$

for some attenuation function f of the distance d between the object and the viewer.

In current good real time renderings, the illumination model used is most often some variant of the one proposed by Phong [8]:

$$I(\lambda) = k_a(\lambda) + k_d(\lambda) <\mathbf{n},\mathbf{l}> + k_s(<\mathbf{e},\mathbf{n}>)^n.$$

Here $k_a(\lambda)$, $k_d(\lambda)$, and k_s denote the coefficients for ambient, diffuse, and specular reflection, respectively, and the exponent n is usually an integer between 0 and 255. Further $\mathbf{n}$ denotes the surface unit normal, $\mathbf{l}$ the unit vector from surface to light, and $\mathbf{e}$ the unit vector from surface to eye, respectively. Moreover, the light source as well as the eye are often assumed to be at infinite distance from the object, which implies that both $\mathbf{l}$ and $\mathbf{e}$ are *constant* vectors.

Since the coefficient k_s is independent of wavelength, highlights are always white. This gives one way to separate the effect of desired modifications to the image intensity introduced by a user into a change of $<\mathbf{n},\mathbf{l}>$ on one hand and $<\mathbf{n},\mathbf{e}>$ on the other hand. Editing the rendered picture in this way prescribes $<\mathbf{n},\mathbf{l}>$ or $<\mathbf{n},\mathbf{e}>$.

6.4.2. Definition of the Overloaded Surface.

To a surface Γ in $\mathbb{R}^3$, given by $\mathbf{r} : D \to \mathbb{R}^3$, $D \subset \mathbb{R}^2$, we associate a new surface $\tilde{\Gamma}$, given by $\mathbf{s} : D \to \mathbb{R}^3$, where

$$\mathbf{s}(u,v) = \mathbf{r}(u,v) + g(u,v) \cdot \mathbf{n}(u,v), \ (u,v) \in D.$$

Here, g is an ordinary real valued function over D and $\mathbf{n}$ is the unit normal to Γ.

To avoid burdening the reader with technical details, we will assume that both Γ and g are of class C^2. We will also assume that the absolute value of g is so small compared with the smallest radius of curvature of Γ that no self-intersection or other anomalies may occur. The resulting surface $\tilde{\Gamma}$, which we will call the *overloaded surface*, will then be of class C^1.

Surfaces of this type have proven to be suitable for several applications within CAD, (see, e.g., [2], [5]) and within *Computer Graphics*, see, e.g., [3]. When g is constant, $\tilde{\Gamma}$ is of course nothing but the offset surface at distance $|g|$ from Γ.

A surface description of this kind has an appealing intuitive feature that should be quite useful in many practical situations. Conceptually, the function g is the mathematical counterpart to the way desired surface position modifications are generally described. Thus, we could expect that a designer working with surfaces should be able to give a fairly precise description of g, e.g., by indicating a suitably chosen collection of level sets for g.

Combined with desired overall properties of g, like convexity or monotonicity, the operator may easily supply the limited information required to enable a suitably designed system to define a rough approximation of g. Of course, the accuracy could easily be improved simply by iterating the procedure.

Besides being in accord with intuition, surface modifications along these lines also have the advantage of releasing the operator from the usual struggle with patches. Often, large modifications of piecewise defined surfaces make a new subdivision necessary. In the procedure just sketched, required large surface variations are reflected by large gradients of g, which could be resolved by an automatic refinement during the approximation.

Since our main interest for this intermediate surface description is to fit a given normal field rather than positions, we will not push this application further, but turn to the computation of the normals to $\tilde{\Gamma}$.

Since $\mathbf{s}'_u = \mathbf{r}'_u + g'_u \cdot \mathbf{n} + g \cdot \mathbf{n}'_u$ and $\mathbf{s}'_v = \mathbf{r}'_v + g'_v \cdot \mathbf{n} + g \cdot \mathbf{n}'_v$ and $\mathbf{n} \times \mathbf{n} = \mathbf{0}$, we get

$$
\begin{aligned}
\mathbf{s}'_u \times \mathbf{s}'_v = \ & \mathbf{r}'_u \times \mathbf{r}'_v + g'_v \cdot \mathbf{r}'_u \times \mathbf{n} - g'_u \cdot \mathbf{r}'_v \times \mathbf{n} + g \cdot (\mathbf{r}'_u \times \mathbf{n}'_v - \\
& \mathbf{r}'_v \times \mathbf{n}'_u) + g \cdot (g'_u \cdot \mathbf{n} \times \mathbf{n}'_v - g'_v \cdot \mathbf{n} \times \mathbf{n}'_u) + g^2 \cdot \mathbf{n}'_u \times \mathbf{n}'_v.
\end{aligned}
$$

From $\|\mathbf{n}\|^2 = 1$ follows $\mathbf{n}'_u \perp \mathbf{n}$ and $\mathbf{n}'_v \perp \mathbf{n}$ so both $\mathbf{n}'_u$ and $\mathbf{n}'_v$ are in the tangent plane to Γ. If E, F, G are the coefficients for the first fundamental form and L, M, N are the coefficients for the second fundamental form, by easy computation, we can write $\mathbf{n}'_u$ and $\mathbf{n}'_v$ in the basis $\mathbf{r}'_u$, $\mathbf{r}'_v$ by the following relations, usually known as Weingarten's equations (see, e.g., [13], [12])

$$\mathbf{n}'_u = (EG - F^2)^{-1} [(FM - GL)\mathbf{r}'_u + (FL - EM)\mathbf{r}'_v]$$

and

$$\mathbf{n}'_v = (EG - F^2)^{-1} [(FN - GM)\mathbf{r}'_u + (FM - EN)\mathbf{r}'_v].$$

Using this and the fact that $\mathbf{r}'_u \times \mathbf{r}'_v = \sqrt{EG - F^2} \cdot \mathbf{n}$, by straightforward computation it follows that

$$\mathbf{r}'_u \times \mathbf{n}'_v - \mathbf{r}'_v \times \mathbf{n}'_u = \frac{2FM - EN - GL}{\sqrt{EG - F^2}} \cdot \mathbf{n}$$

and

$$\mathbf{n}'_u \times \mathbf{n}'_v = \frac{LN - M^2}{\sqrt{EG - F^2}} \cdot \mathbf{n}.$$

Recalling that the mean curvature H and the Gaussian curvature K are expressed through the coefficients of the first and second fundamental forms as

$$H = \frac{2FM - EN - GL}{EG - F^2}$$

and

$$K = \frac{LN - M^2}{EG - F^2},$$

we may write

$$\mathbf{r}'_u \times \mathbf{n}'_v - \mathbf{r}'_v \times \mathbf{n}'_u = -2\sqrt{EG - F^2} \cdot H \cdot \mathbf{n},$$

$$\mathbf{n}'_u \times \mathbf{n}'_v = \sqrt{EG - F^2} \cdot K \cdot \mathbf{n}.$$

Hence

$$\mathbf{s}'_u \times \mathbf{s}'_v = $$
$$[(g'_v \mathbf{r}'_u \times \mathbf{n} - g'_u \mathbf{r}'_v \times \mathbf{n}) + g(g'_v \mathbf{n}'_u - g'_u \mathbf{n}'_v) \times \mathbf{n}] +$$
$$\sqrt{EG - F^2}(1 - 2gH + g^2 K) \cdot \mathbf{n},$$

which is splitting of the normal $\mathbf{s}'_u \times \mathbf{s}'_v$ for $\tilde{\Gamma}$ into one component in the tangent plane to Γ and one component along the normal $\mathbf{n}$ to Γ.

6.4.3. The Linearized Normal. The explicit representation for $\mathbf{s}'_u \times \mathbf{s}'_v$ obtained in the preceding section shows that it depends on g, g'_u, and g'_v in a nonlinear way. Since our main objective is to develop efficient procedures to find a surface whose normal field is a sufficiently good approximation of a given unit vector field, we would prefer to approximate the problem with linear ones.

In this section, we will show that the normals to Γ admit good approximations that are linear in g, g'_u, and g'_v. To be more precise, we will approximate the normal $\frac{1}{\sqrt{EG-F^2}} \cdot \mathbf{s}'_u \times \mathbf{s}'_v$ to $\tilde{\Gamma}$ with one of the following vectors

$$\mathbf{m} = \frac{1}{\sin\beta} \left(\frac{g'_v}{\|\mathbf{r}'_v\|} \cdot \hat{\mathbf{r}}'_u - \frac{g'_u}{\|\mathbf{r}'_u\|} \cdot \hat{\mathbf{r}}'_v \right) + (1 - 2gH) \cdot \mathbf{n}$$

or

$$\overline{\mathbf{m}} = \frac{1}{\sin\beta} \left(\frac{g'_v}{\|\mathbf{r}'_v\|} \cdot \hat{\mathbf{r}}'_u - \frac{g'_u}{\|\mathbf{r}'_u\|} \cdot \hat{\mathbf{r}}'_v \right) + \mathbf{n}.$$

Here β denotes the angle between $\mathbf{r}'_u$ and $\mathbf{r}'_v$. Further, $\hat{\mathbf{r}}'_u = \frac{\mathbf{r}'_u \times \mathbf{n}}{\|\mathbf{r}'_u\|}$ and $\hat{\mathbf{r}}'_v = \frac{\mathbf{r}'_v \times \mathbf{n}}{\|\mathbf{r}'_v\|}$.

Since $\mathbf{n}$ is orthogonal to $\mathbf{r}'_u$ and $\mathbf{r}'_v$, both $\hat{\mathbf{r}}'_u$ and $\hat{\mathbf{r}}'_v$ are unit vectors in the tangent plane. Just by comparing $\frac{1}{\sqrt{EG-F^2}} \cdot \mathbf{s}'_u \times \mathbf{s}'_v$ to $\mathbf{m}$, it turns out that we must estimate the term

$$\frac{g(g'_v \mathbf{n}'_u - g'_u \mathbf{n}'_v) \times \mathbf{n}}{\sin\beta \|\mathbf{r}'_u\| \|\mathbf{r}'_v\|}.$$

This will be done in the following lemma.

LEMMA 6.4.1. *Let k_1 and k_2 be the maximum and minimum curvature to Γ, respectively, and let ϕ be the angle between the vector*

$$\mathbf{v} = -g'_v \cdot \mathbf{r}'_u + g'_u \cdot \mathbf{r}'_v$$

and the direction of maximum curvature to Γ. Then

$$\left\| \frac{g(g'_v \mathbf{n}'_u - g'_u \mathbf{n}'_v) \times \mathbf{n}}{\sin\beta \|\mathbf{r}'_u\| \|\mathbf{r}'_v\|} \right\|^2 = \frac{g^2}{\sin^2\beta} \|\mathbf{u}\|^2 (k_1{}^2 \cos^2\phi + k_2{}^2 \sin^2\phi),$$

with

$$\mathbf{u} = \frac{g'_v}{\|\mathbf{r}'_v\|} \cdot \hat{\mathbf{r}}'_u - \frac{g'_u}{\|\mathbf{r}'_u\|} \cdot \hat{\mathbf{r}}'_v.$$

Proof. Expanding $\|g'_u \mathbf{n}'_v - g'_v \mathbf{n}'_u\|^2$, we get

$$\|g'_u \mathbf{n}'_v - g'_v \mathbf{n}'_u\|^2 =$$
$$(g'_u)^2 \|\mathbf{n}'_v\|^2 - 2g'_u g'_v <\mathbf{n}'_u, \mathbf{n}'_v> + (g'_v)^2 \|\mathbf{n}'_u\|^2.$$

By simple direct computations from Weingarten's equations for $\mathbf{n}'_u$ and $\mathbf{n}'_v$ in the preceeding section, we obtain

$$\|\mathbf{n}'_u\|^2 = 2LH - EK$$

$$< \mathbf{n}'_u, \mathbf{n}'_v > = 2MH - FK$$

$$\|\mathbf{n}'_v\|^2 = 2NH - GK.$$

Thus

$$\|g'_u \mathbf{n}'_v - g'_v \mathbf{n}'_u\|^2$$
$$= (g'_u)^2(2NH - GK) - 2g'_u g'_v(2MH - FK) + (g'_v)^2(2LH - EK)$$
$$= [(g'_u)^2 N - 2g'_u g'_v M + (g'_v)^2 L] \cdot 2H$$
$$\quad - [(g'_u)^2 G - 2g'_u g'_v F + (g'_v)^2 E] \cdot K.$$

Recalling (see, e.g., [11]) that the normal curvature to Γ in the direction given by the vector $\mathbf{v} = -g'_v \cdot \mathbf{r}'_u + g'_u \cdot \mathbf{r}'_v$ is

$$k(-g'_v, g'_u) = \frac{(-g'_v)^2 L + 2(-g'_v)g'_u M + (g'_u)^2 N}{(-g'_v)^2 E + 2(-g'_v)g'_u F + (g'_u)^2 G}$$

we can write

$$\|g'_u \mathbf{n}'_v - g'_v \mathbf{n}'_u\|^2 =$$
$$[(-g'_v)^2 E + 2(-g'_v)g'_u F + (g'_u)^2 G] [2k(-g'_v, g'_u)H - K].$$

From the definition of $\hat{\mathbf{r}}'_u$ and $\hat{\mathbf{r}}'_v$, it is immediate that the first factor equals $\|\mathbf{u}\|^2 \|\mathbf{r}'_u\|^2 \|\mathbf{r}'_v\|^2$. By Euler's formula (see, e.g., [11]),

$$k(-g'_v, g'_u) = k_1 \cos^2\phi + k_2 \sin^2\phi$$

and since $H = \frac{1}{2}(k_1 + k_2)$ and $K = k_1 k_2$,

$$2k(-g'_v, g'_u)H - K =$$
$$(k_1 \cos^2\phi + k_2 \sin^2\phi)(k_1 + k_2) - k_1 k_2 = (k_1^2 \cos^2\phi + k_2^2 \sin^2\phi)$$

and the lemma is proved.

Summing up the discussion on the normal and its linearizations $\mathbf{m}$ and $\overline{\mathbf{m}}$, we have the following estimation for the error caused by replacing the normal by one of the linearizations.

THEOREM 6.4.1. *Let k_1 and k_2 be the maximum and minimum curvature to Γ, respectively, and let ϕ be the angle between the vector*

$$\mathbf{u} = \frac{g'_v}{\|\mathbf{r}'_v\|} \cdot \hat{\mathbf{r}}'_u - \frac{g'_u}{\|\mathbf{r}'_u\|} \cdot \hat{\mathbf{r}}'_v$$

and the direction of minimum curvature to Γ. If

$$\mathbf{m} = \frac{1}{\sin\beta}\left(\frac{g'_v}{\|\mathbf{r}'_v\|} \cdot \hat{\mathbf{r}}'_u - \frac{g'_u}{\|\mathbf{r}'_u\|} \cdot \hat{\mathbf{r}}'_v\right) + (1 - 2gH) \cdot \mathbf{n}$$

and

$$\overline{\mathbf{m}} = \frac{1}{\sin\beta}\left(\frac{g'_v}{\|\mathbf{r}'_v\|}\cdot\hat{\mathbf{r}}'_u - \frac{g'_u}{\|\mathbf{r}'_u\|}\cdot\hat{\mathbf{r}}'_v\right) + \mathbf{n},$$

then

$$\left\|\frac{\mathbf{s}'_u\times\mathbf{s}'_v}{\sqrt{EG-F^2}} - \mathbf{m}\right\|^2 = \frac{g^2}{\sin^2\beta}\|\mathbf{u}\|^2(k_1{}^2\cos^2\phi + k_2{}^2\sin^2\phi) + g^4K^2$$

and

$$\left\|\frac{\mathbf{s}'_u\times\mathbf{s}'_v}{\sqrt{EG-F^2}} - \overline{\mathbf{m}}\right\|^2 =$$
$$\frac{g^2}{\sin^2\beta}\|\mathbf{u}\|^2(k_1{}^2\cos^2\phi + k_2{}^2\sin^2\phi) + (-2gH + g^2K)^2.$$

Proof. The theorem follows directly from the lemma since, as is readily seen by a direct computation, the vectors $\mathbf{u}$ and $\mathbf{v}$ in the lemma are orthogonal and from the fact that the principal directions are also orthogonal.

REMARK 6.4.1. *From the theorem we may immediately draw two conclusions:*

- *For the fairly flat surfaces that are so common among the styling dependent surfaces in the automotive industry, the approximation is quite good irrespectively of any (reasonable) size of $|g|$;*

- *For any surface we may approximate as close as we want, just by making $|g|$ small. To make $|g|$ small, all we need to do is essentially to iterate the process.*

6.4.4. The linearized fitting problem.

We now return to the problem mentioned in the beginning of this section. Thus we suppose that there is given a unit vector field $\tilde{\mathbf{n}}$ over Γ and that we wish to perturbate Γ into a surface $\tilde{\Gamma}$, whose unit normal field coincides with $\tilde{\mathbf{n}}$.

From the theorem of the preceding section, we know how to generate linearized problems and also how well their solutions approximate a possible solution to the original problem. However, we have not established that any of these problems do admit a solution. We first discuss this question for the linearized problem and start by giving it a more precise formulation.

Suppose that the domain of definition D for Γ is a simply connected ("without holes") domain in $\mathbb{R}^2$ and that $\tilde{\mathbf{n}} : D \to S^2$, where S^2 denotes the unit sphere in $\mathbb{R}^3$, is of class C^1.

For the setup of the linearized problem, it is convenient to express $\tilde{\mathbf{n}}$ in our local basis $\{\hat{\mathbf{r}}'_u, \hat{\mathbf{r}}'_v, \mathbf{n}\}$, as

$$\tilde{\mathbf{n}} = a\cdot\hat{\mathbf{r}}'_u + b\cdot\hat{\mathbf{r}}'_v + c\cdot\mathbf{n}$$

with a, b, and $c \in C^1(D)$. Now choosing g so that the linearized normal

$$\overline{\mathbf{m}} = \frac{1}{\sin\beta}\left(\frac{g'_v}{\|\mathbf{r}'_v\|}\cdot\hat{\mathbf{r}}'_u - \frac{g'_u}{\|\mathbf{r}'_u\|}\cdot\hat{\mathbf{r}}'_v\right) + \mathbf{n}$$

to $\tilde{\Gamma}$ equals $\frac{1}{c} \cdot \tilde{\mathbf{n}}$ is equivalent to solving the following problem.

Find $g \in C^2(D)$ such that

$$
\begin{cases}
g'_u = p \\
g'_v = q
\end{cases}
$$

with $p = -\frac{b}{c}\|\mathbf{r}'_u\| \sin \beta$ and $q = \frac{a}{c}\|\mathbf{r}'_v\| \sin \beta$.

By a well-known result in calculus, this problem has a solution if and only if $p'_v = q'_u$. *For an arbitrary vector field $\tilde{\mathbf{n}}$, this relation may very well be violated!*

Checking the condition for a given vector field is not easy. It would be natural to inquire if we could simply disregard the condition, given that we ask for an approximate solution. From a practical point of view, this would be enough, if the error could be made so small, that it would not affect the current application. Our first result is in the negative: the error cannot be made arbitrarily small with less than the condition satisfied.

PROPOSITION 6.4.1. *Assume that for each $\epsilon > 0$ there is a function $g \in C^2$ such that over each compact subset C of the convex set D,*

$$
\max |g'_u - p| < \epsilon, \quad \max |g'_v - q| < \epsilon.
$$

Then $p'_v = q'_u$.

Proof. By the hypothesis, we can find a sequence $(g_k)_{k=1}^{\infty}$ in C^2 such that $\frac{\partial g_k}{\partial u} \to p$ and $\frac{\partial g_k}{\partial v} \to q$, uniformly on compact subsets of D, as $k \to \infty$, and such that $g_k(0,0) = 0$ (there is of course no restriction of generality to assume that $(0,0) \in D$). Define f in D by

$$
f(u,v) = \left(\int_0^1 p(tu, tv)dt \right) u + \left(\int_0^1 q(tu, tv)dt \right) v.
$$

Since

$$
\frac{d}{dt}(g_k(tu, tv)) = \frac{\partial g_k}{\partial u}(tu, tv) \cdot u + \frac{\partial g_k}{\partial v}(tu, tv) \cdot v,
$$

and $\int_0^1 \frac{d}{dt}(g_k(tu, tv))dt = g_k(u,v) - g_k(0,0)$, it follows that $g_k \to f$, uniformly on compact subsets of D, as $k \to \infty$. A simple argument (see, e.g., [15], p. 87) then shows that $f'_u = p$ and $f'_v = q$. Since p and q are in C^1, the function f is in C^2, and this proves the proposition.

The result is just qualitative and does not shed any light on the question on how the error may possibly depend on the way the condition is violated. Since the maximum norm, though natural, is not so well adapted for actual computations, we will use the L^2 norm instead.

PROPOSITION 6.4.2. *Assume that g solves the problem*

$$
\min \int \int ((g'_u - p)^2 + (g'_v - q)^2)du\, dv,
$$

where the double integral extends over D. Then g is a solution to the interior Neumann problem

$$\begin{cases} \Delta g & = p'_u + q'_v \ in \ D \\ \frac{\partial g}{\partial n} & = n_1 p + n_2 q \ on \ \partial D \end{cases}.$$

Here, $n = (n_1, n_2)$ is the interior unit normal to ∂D and $\frac{\partial g}{\partial n}$ is the normal derivative to g.

The proposition turns the question into one of the best-known problems in mathematics, the Neumann problem for Poisson's equations. It is interesting to notice that this equation also enters naturally in another way.

Up to now, we have assumed that there are no boundary conditions to be satisfied. Considering the problem, if there is a solution, its differential must equal zero. Thus a solution is unique, apart from that some arbitrary constant may be added, leaving very little space to satisfy a boundary condition.

In several cases, however, boundary conditions must also be taken into account. A very natural requirement, for instance, is that our deformation g should vanish at the boundary of the domain. This is exactly the case when the surface we are modifying should fit together with other surfaces that are already present. In general, if we insist that our modified surface should fit exactly, the best we can do is to search for an approximate solution to the normal fitting problem. Still using the L^2 norm, this leads to another classical boundary value problem, the Dirichlet problem for Poisson's equation.

PROPOSITION 6.4.3. *Assume that g solves the problem*

$$\min \int \int ((g'_u - p)^2 + (g'_v - q)^2) du \ dv,$$

subject to $g = 0$ at ∂D. Then g is a solution to the interior Dirichlet problem

$$\begin{cases} \Delta g & = p'_u + q'_v \ in \ D \\ g & = 0 \ on \ \partial D \end{cases}.$$

The last two propositions follow easily from a result within the calculus of variation (see, e.g., [9], pp. 469–474 and pp. 265–267).

6.4.5. Numerical Experiments. We will now present the outcome of some computations based on the results obtained in the section. In order to compare the results of the approximate computations with an exact result, we have constructed the given field of unit vectors and the basic surface in the following manner.

First, we have selected a rather curved surface patch Γ_0. Indeed, Γ_0 is one the most curved regions of the hood of a Volvo in the 700 series, a small region in the front, situated above the joint between the radiator grill and headlight.

Next, we have deformed Γ_0, leaving its boundary unchanged, into our basic surface Γ. Figure 6.3 shows a rendering of both surfaces. Here, Γ_0 is the one with a concentrated highlight, whereas that of Γ is scattered, like a comet's

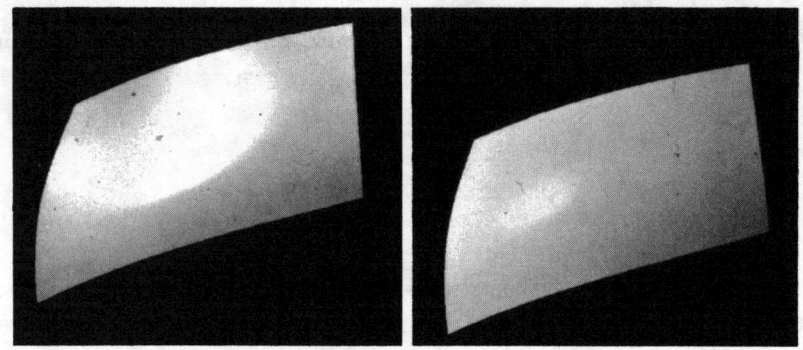

FIG. 6.3. *A rendering of the surfaces* Γ *(left) and* Γ_0 *(right).*

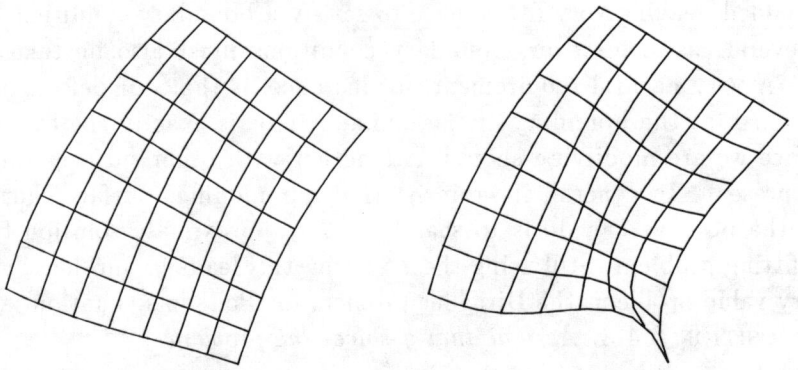

FIG. 6.4. *Offset surfaces at distance* 10 mm *of* Γ_0 *(left) and* Γ *(right).*

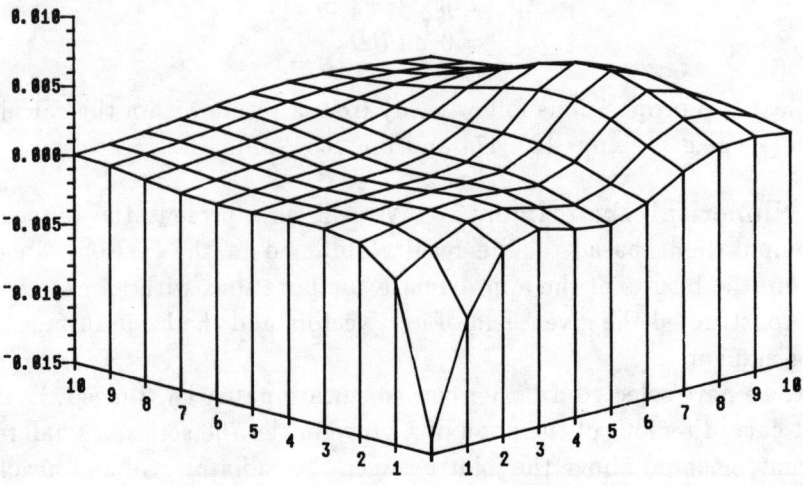

FIG. 6.5. *Gaussian curvature of* Γ.

tail. In Figure 6.4 the offset surfaces at distance 10 mm to both surfaces are shown, from which it immediately turns out that the normal field to Γ is rather extreme. In comparison with usual outer skin surfaces, it is highly curved, the smallest radius of curvature is just 8.2 mm. Figures 6.5 and 6.6 are plots of the Gaussian—and mean—curvature of Γ, respectively.

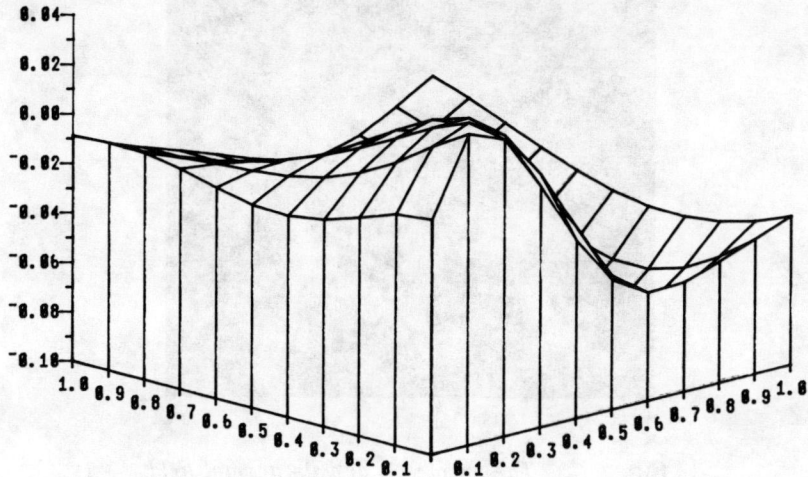

FIG. 6.6. *Mean curvature of Γ.*

The given unit vector field ñ over Γ will now be defined. Chose some fixed viewing direction $\mathbf{w} \in S^2$ and let for each $(u,v) \in D = [0,1]^2$, $l(u,v)$ be the line through the point $\mathbf{r}(u,v) \in \Gamma$ with direction $\mathbf{w}$. Further, let (u_0, v_0) be the parameter value for the point on Γ_0 where $l(u,v)$ intersects Γ_0 and put ñ$(u,v) = \mathbf{n}_0(u_0, v_0)$, where $\mathbf{n}_0$ is the unit normal to Γ_0.

Figures 6.7 and 6.8 show Γ and Γ_0 together with some of their normals and suggest that there should be a fairly large discrepancy between ñ and $\mathbf{n}$. Indeed, the biggest angle between ñ and $\mathbf{n}$ exceeds 17 degrees.

Of the many possible ways to solve the problem approximately, for these experiments we have chosen two very simple ones. For reference to an alternative method, see [4], p. 168.

Since the domain D is rectangular, we divide it into a rectangular grid by subdivisions $0 = u_0 < u_1 < \cdots < u_m = 1$ and $0 = v_0 < v_1 < \cdots < v_m = 1$ and write

$$g_{m,n}(u,v) =$$
$$\sum_{i=0}^{m}\sum_{j=0}^{n} a_{ij}\phi_i(u)\phi_j(v) + \sum_{i=0}^{m}\sum_{j=0}^{n} b_{ij}\psi_i(u)\phi_j(v) +$$
$$\sum_{i=0}^{m}\sum_{j=0}^{n} c_{ij}\phi_i(u)\psi_j(v) + \sum_{i=0}^{m}\sum_{j=0}^{n} d_{ij}\psi_i(u)\psi_j(v)$$

where ϕ_i and ψ_j are the univariate cubic Hermite polynomials associated with these partitions. For a domain with a curved boundary that must be followed

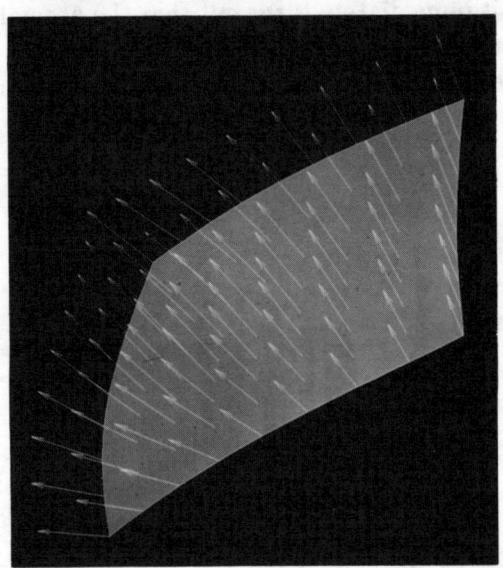

FIG. 6.7.   *The surface* Γ *and its normal field.*

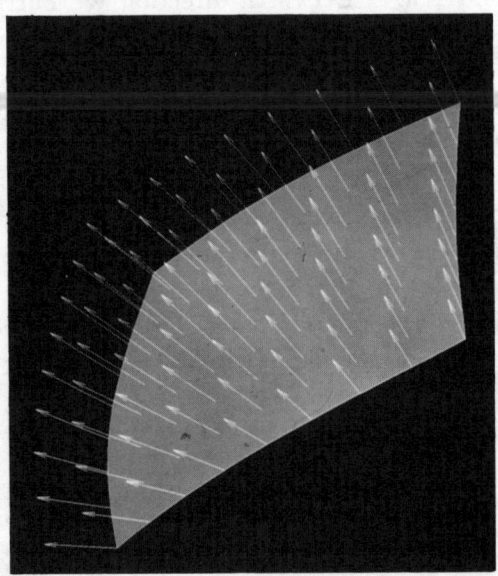

FIG. 6.8.   *The surface* Γ_0 *and its normal field.*

exactly, the elements containing parts of the boundary could be formed by blending (see, e.g., [7]).

In one of the methods, we determine $a_{ij}, b_{ij}, c_{ij}, d_{ij}$ as the least square (LSQ) fit at MN points (u, v), with $M > m$ and $N > n$. In the other, these coefficients are determined as the solution to the LP-problem to minimize ϵ where the absolute value of the difference between $g_{m,n}(u, v)$ and the right-hand side does not exceed ϵ for these MN points.

In all our experiments, the LSQ fit is much faster than the LP method. However, using just two standard subroutines for dense problems, it is hard to draw conclusions on the *methods*. The LP method gives a slightly less maximum error, and has also the advantage to make it very easy to respecify other conditions as well.

Even for rather few variables, the problems are sparse, and we expect that sparse solvers will cut the solution time considerably.

With $m = n = 1$, we get a problem with 16 variables, a maximum angle between ñ and **n** of 3.03 degrees with LSQ and 2.87 with LP, requiring a CPU time of 22 milliseconds on an IBM 3090 for the former and about four times that for the latter.

With $m = n = 2$, the problem has 36 variables, gives a maximum angle between ñ and **n** of 0.74 degrees and LSQ requires 140 milliseconds CPU.

Finally $m = n = 4$ gives a problem with 100 variables, a maximum angle between ñ and **n** of 0.13 degrees and LSQ requires 645 milliseconds of CPU.

In the case with $m = n = 1$, we have also iterated the process, replacing Γ with $\tilde{\Gamma}$. With the same number of variables, this improved accuracy only slightly. However, the maximum difference between the exact and the linearized normal was reduced considerably, from about $5 \cdot 10^{-3}$ to about $5 \cdot 10^{-7}$.

Recalling the estimate for the error due to linearization of the normal, we would anticipate such a fast decay, due to the double effect of reducing the norm of **u** and the absolute value of g.

In any case, despite the high curvature of Γ, the error due to linearization could safely be neglected. It is the extremely bad behavior of our Γ, causing very rapid changes of g and its derivatives, that limits the accuracy for a given subdivision. These properties of the right-hand sides p and q in the problem introduced in the beginning of §6.4.4 are illustrated through their level curves in Fig. 6.9 and Fig. 6.10.

Acknowledgment

The research of B. E. J. Dahlberg was partially supported by AFOSR grant 89-0455 and by ONR grant N00014-90-J-1343.

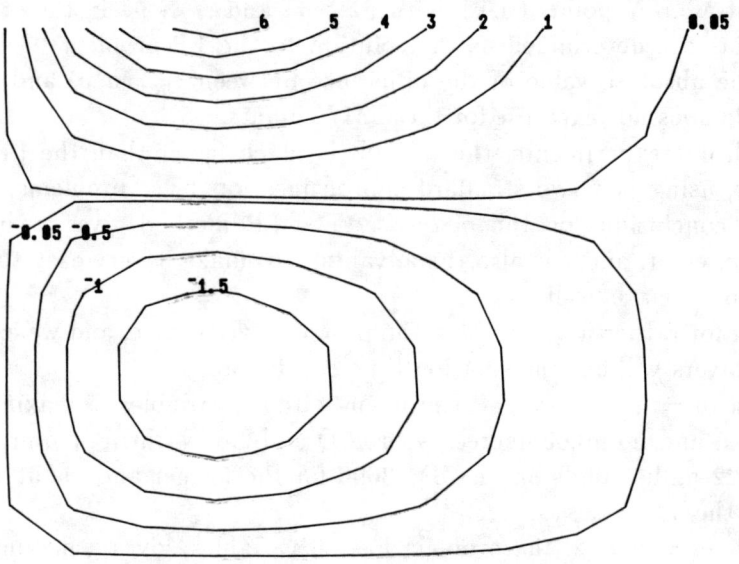

FIG. 6.9.   *Level curves of the function p.*

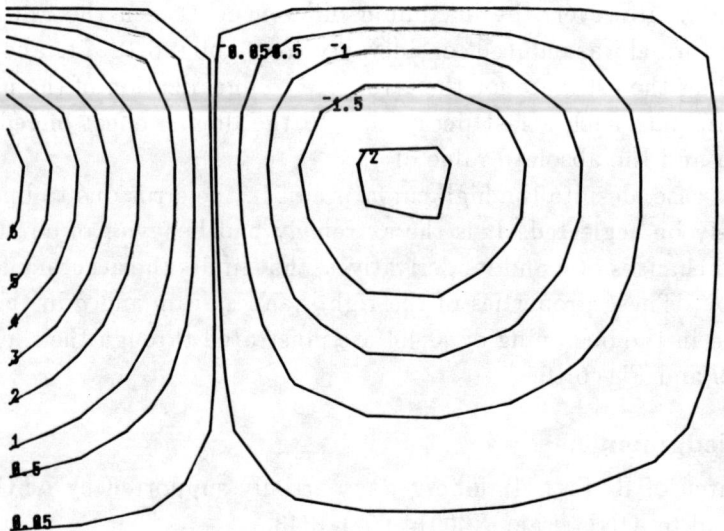

FIG. 6.10.   *Level curves of the function q.*

References

[1] E. Andersson, R. Andersson, M. Boman, T. Elmroth, B. E. J. Dahlberg, and B. Johansson, *Automatic construction of surfaces with prescribed shape*, Comput. Aided Des., 20 (1988), pp. 317–324.

[2] R. E. Barnhill, *Computer aided geometric design*, in Approximation Theory 6, C. K. Chui, L. L. Schumaker, and J. D. Ward, eds., Academic Press Inc, New York, 1990.

[3] J. F. Blinn, *Simulation of wrinkled surfaces*, Proc. SIGGRAPH 78, Computer Graphics, 12 (1978), pp. 286–292.

[4] Chui, C.K., *Multivariate Splines*, Society for Industrial and Applied Mathematics, Philadelphia, 1988.

[5] T. A. Foley, D. A. Lane, G. M. Nielson, R. Franke and H. Hagen, *Interpolation of scattered data on closed surfaces*, Comput. Aided Geom. Des., 7 (1990), pp. 303–312.

[6] H. Folkesson, *CAE in automotive development*, Thesis for Degree of Licentiate of Engineering, Report No 1986-10-01, Division of Machine Elements, Chalmers University of Technology, Sweden, 1986.

[7] H. Hagen, *Geometric modelling of smooth surfaces using triangular patches*, in Theory and Practice of Geometric Modelling, W. Straßer and H-P. Seidel, eds., Springer-Verlag, New York, 1989.

[8] R. Hall, *Color reproduction and illumination models*, in Techniques for Computer Graphics, D. F. Rogers and R. A. Earnshaw, eds., Springer-Verlag, New York, 1987.

[9] B.K.P. Horn,, *Robot Vision*, The MIT Press and McGraw–Hill, Boston and New York, 1986.

[10] L. Hörmander, *The boundary problems of physical geodesy*, Arch. Rat. Mech. Anal., 62 (1976), pp. 1–52.

[11] B. O'Neill, *Elementary Differential Geometry*, Academic Press, New York, 1966.

[12] J.J. Stoker, *Differential Geometry*, Wiley-Interscience, New York, 1969.

[13] D. Struik, *Lectures on Classical Differential Geometry*, Addison–Wesley, Reading, Massachusetts, 1950.

[14] M. Tovey, *Computer-aided vehicle styling*, Comput. Aided Des., 21 (1989), pp. 172–181.

[15] F. Treves, *Topological Vector Spaces, Distributions and Kernals*, Academic Press, New York, 1967.

Exact Conversion of a Trimmed Nonrational Bézier Surface into Composite or Basic Nonrational Bézier Surfaces

A. E. Vries-Baayens and C. H. Seebregts

7.1. Introduction

Quite a number of users of computer aided design (CAD) systems assume that when the international ISO standard STEP[1] becomes available, all problems with the data exchange of product models between dissimilar CAD systems will be solved. In principle this assumption is valid, but in practice a number of problems will occur that have been recognized during the development of the standard and for which no adequate solutions have yet been found. One such problem that will arise is the difference in functionality between different CAD systems, and between CAD systems and the devices for the numerical control of machines such as NC machines. In particular the abstraction level on which a product model is represented can differ between CAD systems. By product model is meant here the wide range of types of information needed for the design and manufacture of a product, of which geometry and topology are parts. Abstraction levels can be used to subdivide a CAD system into subsystems (subsets of the collections of objects forming the system) for which relations between the subsystems can be given simultaneously. On the highest levels a hierarchy can be given of systems, assemblies, parts, or elements (see [31]). On the lowest level the geometry and topology are defined. Between the different abstraction levels transfer links can be given. Shah and Wilson distinguish a number of abstraction links [31]. If information transfer is from a task that operates at a higher level of abstraction to one at a lower level of abstraction, this transformation link is called a *decomposition* link. If information transfer is in the opposite direction it is called a *reconstruction* link. If product data are represented in different forms but at the same level of abstraction, this is called a *mapping* link. When data exchange between CAD systems takes place, a mechanism will be necessary to handle the differences in abstraction levels and to ensure that the data exchange

---

[1]STandard for the Exchange of Productmodel data.

can take place smoothly. This difference in abstraction level will play an increasingly important role and will influence the data exchange process greatly, especially for the new generation CAD systems. Present-day CAD systems do not incorporate abstraction levels to any extent and normally represent the geometry and topology of a product at the lowest level. Solid modelers (B-REP and CSG) are able to handle and store topology whereas this is far more difficult for surface- and wire-frame modelers.

In this paper we will analyze some of the problems with data exchange between CAD systems on the lowest abstraction level and will offer a solution for one of these problems. At present, the following problems exist, which the new standard STEP aims to solve.

1. On the sending CAD system, entities have been defined in ways that are not covered by the standard STEP. These entities therefore can not be translated into entities of the STEP neutral format. This means that the information about these entities will be lost in the data exchange process.

2. The receiving CAD system does not recognize a number of entities defined in the STEP neutral format, or these entities are defined with more than the allowable constraints, which are not included in the permitted set (constraints on degree, parameter distributions, etc.). Therefore, data that describe those entities cannot be received or be interpreted.

3. The functionality of NC machines is more restricted than that of present-day CAD systems, in that most NC machines are only able to handle patch information in B-spline, Bézier, or polynomial representation. So before a STEP file can be sent to such a NC machine, unacceptable entities will have to be converted into an acceptable form.

If data exchange has to take place, a file in the internal format of the sending CAD system has to be mapped into a neutral file format (pre-processing) and the neutral file format has to be mapped into the internal format of the receiving CAD system. In both mappings entities cannot be included in the permitted set and conversions will then be needed. The question is then whether the type of conversions needed can be identified. Therefore, in this paper we develop a classification of the geometry data and topology data of a product model on the basis of which the different types of conversions can be identified. Hereafter, we can then define the subject of our paper and give an overview of the paragraphs.

Suppliers of CAD systems offer a number of tools that can be added to the geometry of a product model to define the topology. The classification of the geometry and the tools offered by a CAD system with which a productmodel can be created, is the following.

1. *Basic geometric entities.* These are the elementary building blocks with which a model can be created. One can distinguish four *types* of entities: points, curves, surfaces, and volumes. Each of these entities can be defined in two-dimensional and/or three-dimensional space. Each entity is explicitly defined by a mathematical expression that completely describes

it. Different types of mathematical descriptions can be used, e.g., parametric, trigoniometric, algebraic. With these, different basic entities can be formed such as straight lines, curves, planes, surfaces, tori, cubes, spheres, conics, free-formed volumes.

2. *Tools for defining the topology that combines a number of entities into a single new entity on a higher level.* This can be done *explicitly*. A collection of basic entities joined end to end together form a nonselfintersecting building block on a higher level than the basic entity level. This is here called a *composite* entity. A composite entity can be repeatedly combined into a composite entity at a higher level.[2] This can also be done *implicitly*. One or more entities on a higher abstraction level are composed out of a collection of basic entities and/or composite entities in combination with Boolean operations. These entities are often called *trimmed* entities if curves or surfaces are used as basic entities because they consist of a basic entity that is trimmed by other basic entities. For volumes, Boolean operations are normally used on entities such as a torus, cube, sphere, or conic.

The data for the geometry of a product model can now be grouped as follows.

1. Data of the basic entities that consist of the mathematical description of the entities and their constraints.

2. Data to join basic entities explicitly into a composite structure. We will call these data "composite data."

3. Data to compose entities implicitly out of basic entities or composite entities by Boolean operations. We will call these data "trimming data."

What cannot be covered by STEP (or any other standard) is a solution for the problem that geometric basic entities and tools for topology are transmitted in forms and with constraints that are not included in the premitted set. The following problems can arise.

1. The receiving CAD system uses different mathematical representations for its basic entities or uses different constraints for the same mathematical representation (lower degree, other parameterization, etc.).

2. The receiving CAD system does not recognize a composite entity structure.

3. The receiving CAD system is not able to recognize implicit defined (trimmed) entities.

---

[2]The standard STEP uses a different terminology for these higher abstraction levels. This terminology depends on the type of modeler (surface, B-Rep, CSG) and on the application area for which the standard is used. The term *composite* entity denotes an explicit combination of curves or surfaces; the term *topology* is used to denote the combination of entities used for boundary representation solid models (vertex, edge, plane, face, shell, object, etc.) while the term *shape* denotes the combination of parts of products on a higher level (assembly-model, etc.) [25].

As yet no solutions have been found. The following conversions are then needed.

1. Conversions of the mathematical representations and constraints.

2. Conversions of composite entities into basic entities.

3. Conversions of trimmed entities into composite or basic entities.

In this paper we will examine the problem how *trimmed surfaces* can be converted into *composite surfaces* or *basic surfaces*, and provide a solution for it. The reason that this problem needs to be solved is that a number of CAD systems and NC machines do not recognize trimmed surfaces. If data have to be transmitted to these systems and machines, these trimmed surfaces have to be converted into explicitly defined surfaces.

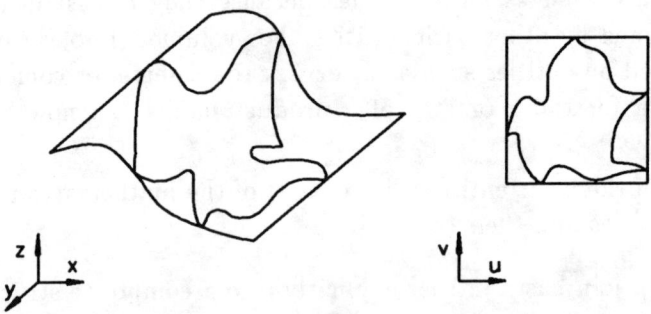

FIG. 7.1. *Trimming of a basic surface by a boundary.*

We first define trimmed surfaces.

Definition. A *trimmed surface* is an implicitly defined surface that consists of a basic surface or a composite surface that is trimmed by one or more boundaries. It is assumed that there is one outer boundary that may enclose zero or more inner boundaries.

Definition. A *boundary* consists of one or more curves resulting from the intersection of one or more surfaces with the basic surface. The boundary will usually be an approximation of these intersections and will partly lie in one surface and partly in the other. We assume here that a boundary lies completely in the basic surface.

In fact, the conversion of a trimmed surface into basic surfaces or composite surfaces requests an explicit description of the area enclosed by the curves of the boundary. We will call this surface the "resulting surface(s)."

Section 7.2 further limits the problem. Section 7.3 gives the state of the art. Section 7.4 deals with the conditions required for a solution for this conversion. Section 7.5 gives an outline of the solution. Section 7.6 discusses the methods used for this solution and the problems that arose with these methods, as well as the solutions found for the problem. Section 7.7 gives a complete description of the algorithm. Section 7.8 gives some examples, and §7.9 discusses unsolved problems.

7.2. Description of the Problem

As stated in §7.1, the problem to be discussed is the conversion of a trimmed (implicitly defined) free-formed surface into composite (explicitly defined) surface or basic surfaces, the so-called "resulting surface(s)." Such a conversion can be exact or approximate. With exact is meant that the geometry of the "resulting surfaces" is precisely the same as the geometry of the basic surface between the boundary. It means that the total outer boundary of the "resulting surfaces" will be an approximation because this boundary was already an approximation of the intersection of one or more surfaces with the basic surface. The area enclosed by the boundary has the same geometry as the original surface. With an approximation is meant that the geometry of the surface enclosed by the boundary is also approximated. Our choice has been to try to find an exact conversion and if this is not possible to look for a good approximation. This conversion can be effected for quite a number of mathematical representations. In this paper the conversion of a trimmed nonrational Bézier surface into composite or basic nonrational Bézier surfaces will be given. The Bézier representation is chosen because it is often used in surface modeling and is computationally stable [12]. The problem that should be solved can now be defined as follows.

Given a single nonrational Bézier basic surface,

$$\mathbf{r}(u,v) = \begin{pmatrix} X \\ Y \\ Z \end{pmatrix} = \sum_{i=0}^{m} \sum_{j=0}^{n} \mathbf{P}_{ij} B_i^m(u) B_j^n(v),$$

where $\quad B_i^m(u) = \binom{m}{i}(1-u)^{m-i} u^i$ for $u \in [0,1)$,

(7.1)

$$B_j^n(v) = \binom{n}{j}(1-v)^{n-j} v^j \text{ for } v \in [0,1),$$

$\mathbf{P}_{ij} = $ the control points of $\mathbf{r}(u,v)$,

$m, n = $ the polynomial degrees for u, respectively, v direction,

and x nonrational Bézier curves for one boundary in three-dimensional space, that *lie on* the Bézier basic surface:

$$\mathbf{c}_k(t) = \begin{pmatrix} X \\ Y \\ Z \end{pmatrix} = \sum_{i=0}^{N(k)} \mathbf{Q}_{k,l} M_l^{N(k)}(t),$$

(7.2)

where $\quad M_l^{N(k)} = \binom{N(k)}{l}(1-t)^{N(k)-l} t^l$ for $t \in [0,1)$

for $\qquad k = 1, \cdots, x$

where k gives the number of the curve of the boundary,

$N(k) = $ the degree of the curve k of the boundary,

$\mathbf{Q}_{k,l} = $ the Bézier control points of $\mathbf{c}_k(t)$.

We seek the explicit definition of the surface(s) within the boundary $c_k(t)$, the so-called resulting surface(s), which must be given in the nonrational Bézier representation.

7.3. State of the Art

Contributions on the subject of trimmed surfaces come mainly from the solid modeling area where the issue of the so-called "sculptured solids" (solids with a free-formed outer surface) is becoming of increasing importance. Most contributions concentrate on the intersection problem of sculptured solids. Casale [5] gives a method that supports trimmed cubic parametric outerfaces of a B-REP model and an outline of a set operation algorithm (see also [6]). The domain of the trimmed outerfaces is represented by a two-dimensional CSG tree. The most important problems are caused by the intersection algorithms for parametric surfaces. Farouki [12] proposes a trimmed surface formulation appropriate to the Boolean combination of primitives bounded by a family of elementary surface patches with dual parametric rational polynomial and implicit algebraic equations.

From the computer graphics area, some contributions on trimmed surfaces are also known. These contributions concentrate on the high-quality rendering methods for trimmed parametric surfaces. Rockwood gives a fast facetting scheme for NURBS surfaces [29], [28]. He partitions the interior of trimmed patches with a uniformly spaced rectangular grid and stitches the grid to the boundary curves with a set of triangles.

In the area of data exchange between (dissimilar) CAD systems a contribution is given by Hoschek on an approximate conversion of trimmed surfaces [17]. He reduces the degree of a NURBS basic surface together with the NURBS curves of the boundary on the basic surface. First the NURBS basic surface of arbitrary degree is reduced to bicubic or biquintic Bézier patches after which the NURBS curves of the boundary are approximated by nonrational B-spline curves. Next the NURBS curves of the boundary are projected on the approximated Bézier patches after which these curves are approximated by the available nonrational B-spline curves by minimizing the absolute error sum of the error vectors. Another contribution comes from Kelder [19], [20] who describes the trimming of one Bézier surface by one boundary consisting of four Bézier curves. He uses a bilinear Coons interpolation method to interpolate the area enclosed by the boundary in the u, v-space of the Bézier surface. For this interpolation he converts the Bézier curves of the boundary and the Bézier basic surface into the polynomial representation. Then he inserts the result of the interpolation in the polynomial description of the basic surface. The end result is converted back into the Bézier representation.

As can be seen from the above, only the method developed by Kelder gives an explicit definition of the area between the boundary curves. The method developed by Kelder can only be used for a boundary consisting of four Bézier boundary curves. So, a more general method to convert the trimmed patch entity into the basic or composite entity must be found.

7.4. Conditions for the Suggested Solution

The solution which will be given holds under the following conditions.

1. The explicit definition of the area enclosed by the boundary should result in surfaces that lie completely *in* the original basic surface. Therefore, the conversion should be exact. The "resulting surface(s)" will be then at least C^0-continuous at the boundary and have internally the same continuity as the original basic surface.

2. The "resulting surface(s)" should all have the nonrational Bézier representation because it should be possible exchange data between present-day dissimilar CAD systems.

3. The "resulting surface(s)" should not have a degree higher than 20 because this is the maximum degree an average CAD system can handle.

4. The number of "resulting surfaces" should be as small as possible.

7.5. Suggested Solution

The resulting surfaces should lie *exactly* in the original basic surface $\mathbf{r}(u, v)$ and be in the nonrational Bézier representation. This can only be achieved if the interpolation of the area enclosed by the boundary takes place in the parameter space. The result of this interpolation will be a description of a two-dimensional area in the u, v-space. Next, this description can be substituted in the description of the basic surface (see [19]). This will result in a definition of the "resulting surface" which will be enclosed by the boundary and will also lie exactly in the basic surface in three-dimensional space. Therefore, it is necessary that the control points of the Bézier curves of the boundary are defined in the u, v-parameter space of basic surface $\mathbf{r}(u, v)$. These control points have to be defined within the parameter range $u \in [0, 1]$ and $v \in [0, 1]$ of the basic surface $\mathbf{r}(u, v)$. The determination of the u, v-values for the control points of these Bézier curves is a separate subject which will not be considered here.

In the standard STEP [10], an entity called "PCURVE" (parameter curve) is defined which has control points given in the parametric space. It is assumed in our algorithm that u, v-values of the control points of the Bézier curves of a boundary are known. The curves i of a boundary can now be defined as follows.

$$\mathbf{c}_k(w) = \begin{pmatrix} u_k(w) \\ v_k(w) \end{pmatrix} = \sum_{l=0}^{N(k)} \mathbf{R}_{k,l} M_l^{N(k)}(w) \text{ for } w \in [0, 1),$$

(7.3)

$\qquad k = 1, \cdots x$ is the number of the boundary curve k,

$\qquad x = $ total number of boundary curves,

$\qquad N(k) = $ the degree of the curve k of the boundary,

$$\mathbf{R}_{k,l} = \begin{pmatrix} r_{u,k,l} \\ r_{v,k,l} \end{pmatrix} = \text{the Bézier control points defined for the } u, v\text{-space,}$$

$$M_l^{N(k)}(v) = \begin{pmatrix} N(k) \\ l \end{pmatrix} (1-w)^{N(k)} w^l \text{ for } w \in [0,1).$$

If one boundary is given, the problem defined in §7.1 can now be divided into two subproblems.

1. Find the explicit description of the area enclosed by x two-dimensional Bézier curves with control points defined in the u, v-space.

2. Find the corresponding definition of this two-dimensional area in the three-dimensional Cartesian space.

In three-dimensional space the area enclosed by three or four curves is usually determined by an interpolation method like the (degenerated) Coons interpolation. So, it seems an obvious choice to use such a method for the first part of this two-dimensional problem. We did choose the linear Coons interpolation method for our solution. Coons' method can interpolate three or four curves. If more than four boundary curves are given, it is necessary to find a method that results in sets of three or four boundary curves each enclosing an area that can then be interpolated. Together these areas should describe the complete area between the boundary curves. In fact, a kind of two-dimensional "triangulation" method for curves should be found. If less than three boundary curves are given, the boundary curves need to be split until three or four curves are obtained.

If the explicit mathematical description of the area between the boundary in u, v-space is found, this description is inserted in the description of the basic surface in order to obtain the corresponding definition in Cartesian space which is the so-called "resulting surface."

The idea to find a solution is now:

1. Store the control points of successive Bézier curves of the boundary counterclockwise.

2. Determine how many curves for a boundary are given.

   a. If fewer than three curves are available, split the curve(s) until three boundary curves are obtained.

   b. If more than four curves are given, try to find sets of curves and connection lines that enclose an area. Together these sets of curves define the complete area between the boundary curves.

3. Convert the Bézier representation of the curves of the boundary and of the basic surface into the polynomial representation:

$$\mathbf{r}(u,v) = \begin{pmatrix} X \\ Y \\ Z \end{pmatrix} = \sum_{i=0}^{m} \sum_{j=0}^{n} \mathbf{p}_{i,j} u^i v^j,$$

(7.4)

$$\mathbf{c}_k(w) = \begin{pmatrix} u_k(w) \\ v_k(w) \end{pmatrix} = \sum_{i=0}^{N(k)} \mathbf{r}_{k,l} w^l,$$

for $u \in [0,1), v \in [0,1), w \in [0,1)$.

4. Carry out the bilinear Coons interpolation, which will result in a polynomial surface representation for each set of three or four curves:

$$\mathbf{r}(s,t) = \begin{pmatrix} u \\ v \end{pmatrix} = \sum_{f=0}^{MS} \sum_{g=0}^{MT} \mathbf{b}_{f,g} s^f t^g,$$

(7.5)

where $s \in [0,1), t \in [0,1)$,

$$\mathbf{b}_{f,g} = \begin{pmatrix} b_{u,f,g} \\ b_{v,f,g} \end{pmatrix}.$$

5. Substitute the mathematical description of the area(s) enclosed by the boundary into the description of the basic surface to get the "resulting surface(s)":

$$(7.6) \quad \mathbf{r}(u,v) = \begin{pmatrix} X \\ Y \\ Z \end{pmatrix} = \sum_{i=0}^{m} \sum_{j=0}^{n} \mathbf{p}_{i,j} \left(\sum_{f=0}^{MS} \sum_{g=0}^{MT} b_{u,f,g} s^f t^g \right)^i \left(\sum_{f=0}^{MS} \sum_{g=0}^{MT} b_{v,f,g} s^f t^g \right)^j.$$

6. Determine the new coefficients for each "resulting surface."

7. Convert the "resulting surface(s)" into the Bézier representation.

It is possible to interpolate each set of three or four *Bézier* curves of the boundary with the help of Coons' method. The description of such a "resulting surface" $\mathbf{r}(s,t)$ in the Bézier representation can be filled in the description of the Bézier basic surface $\mathbf{r}(u,v)$ as is done for the polynomial case in (7.6). However, we had problems in finding a general way to determine the Bézier control points of each "resulting surface." The conversions to and from the polynomial representation will make our method less accurate (see [12]–[14]).

As we will see later, this simple proposal to find the explicit definition of the area enclosed by the boundary turns out to be complex in practice. In the next section, the methods and existing algorithms we used will be given. Hereafter, a detailed description of the solution is given.

7.6. Methods Used for the Solution

For our solution we used a number of algorithms and methods developed in the field of computational geometry and geometric modeling, namely,

1. To split Bézier curves of a boundary if fewer than three curves are given. At most $M - 1$ knots have to be inserted for a certain value of t (see (7.3)). We

use the algorithm developed by Boehm for the insertion of one knot per turn in a B-spline curve after which the new control points are calculated again (see [1]–[4]).

2. Conversion from the Bézier representation into the polynomial representation is executed with methods described in [7] and [12].

3. For the Coons interpolation, each pair of opposite curves should have the same degree. The Bézier boundary curve with the lower degree is elevated using the method described by Farin [11, p. 47].

4. To interpolate three or four curves, the bilinear Coons interpolation is used [8], [9]. This method has been developed to interpolate arbitrary three-dimensional curves for quadrilateral elements with three lateral elements as a degenerate case.

5. If more than four Bézier boundary curves are available we developed a method which we called the "quadrangulation" method to "quadrangulate" these curves.

The bilinear Coons interpolation method and the "quadrangulation" method will be discussed in the following subsections in greater detail. The Coons interpolation method is discussed because it can cause a great number of problems when used in the (two-dimensional) u, v-space. We will discuss methods to detect these problems and also give solutions. The reason to use the Coons interpolation method in the u, v-space is that the result of this interpolation can be obtained in the polynomial, Bézier, or B-spline representation depending on the representation of the boundary. The result of the interpolation can then be inserted in the original Bézier representation of the surface and used by present-day CAD systems (see the conditions in §7.4).

7.6.1. Linear Coons Interpolation.
Consider a surface patch $\mathbf{P}(s,t)$ enclosed by four Bézier curves where the parameters $s \in [0,1)$ and $t \in [0,1)$ are parameters defining the surface. Let $\mathbf{s}(0,0)$, $\mathbf{s}(1,0)$, $\mathbf{s}(0,1)$, and $\mathbf{s}(1,1)$ be the position vectors at the four corner points of the patch and denote the four curves of a boundary by $\mathbf{s}(s,0)$, $\mathbf{s}(s,1)$, $\mathbf{s}(0,t)$, and $\mathbf{s}(1,t)$. The well-known bilinear Coons surface is given as follows [8], [9]:

$$
\begin{aligned}
\mathbf{P}(s,t) = {} & (1-s)\mathbf{s}(0,t) + (s)\mathbf{s}(1,t) + (1-t)\mathbf{s}(s,0) + \\
& (t)\mathbf{s}(s,1) - (1-s)(1-t)\mathbf{s}(0,0) - \\
& (1-s)(t)\mathbf{s}(0,1) - (s)(1-t)\mathbf{s}(1,0) - \\
& (s)(t)\mathbf{s}(1,1).
\end{aligned}
\tag{7.7}
$$

Two curves that are lying opposite to each other are blended with this formula by straight lines (a ruled surface). The surface patch formula (7.4) can be decomposed into three terms that are all expressions in terms of s and t. Thus, each term represents a surface that can be represented as $\mathbf{P}_a(s,t)$, $\mathbf{P}_b(s,t)$, and $\mathbf{P}_c(s,t)$ where (see [33])

(7.8)
$$\begin{aligned}
\mathbf{P}_a(s,t) &= (1-s)\mathbf{s}(0,t) + (s)\mathbf{s}(1,t), \\
\mathbf{P}_b(s,t) &= (1-t)\mathbf{s}(s,0) + (t)\mathbf{s}(s,1), \\
\mathbf{P}_c(s,t) &= (1-s)(1-t)\mathbf{s}(0,0) + (1-s)(t)\mathbf{s}(0,1) \\
&\quad + (s)(1-t)\mathbf{s}(1,t) + (s)(t)\mathbf{s}(1,1).
\end{aligned}$$

Thus the Coons surface patch can be interpreted as being composed of three surfaces $\mathbf{P}_a(s,t)$, $\mathbf{P}_b(s,t)$, $\mathbf{P}_c(s,t)$ where

(7.9) $\qquad \mathbf{P}(s,t) = \mathbf{P}_a(s,t) + \mathbf{P}_b(s,t) - \mathbf{P}_c(s,t)$. (See Fig. 7.2.)

When surface $\mathbf{P}_b(s,t)$ is added to surface $\mathbf{P}_a(s,t)$, the curves corresponding to $s = 0$, $s = 1$, $t = 0$, and $t = 1$ include some excess curves in addition to the boundary curves $\mathbf{s}(0,t)$, $\mathbf{s}(1,t)$, $\mathbf{s}(s,0)$, and $\mathbf{s}(s,1)$. These excess curves are

(7.10)
$$\begin{aligned}
\mathbf{R}_A(s,0) &= (1-s)\mathbf{s}(0,0) + (s)\mathbf{s}(1,0), \\
\mathbf{R}_B(s,1) &= (1-s)\mathbf{s}(0,1) + (s)\mathbf{s}(1,1), \\
\mathbf{R}_B(0,t) &= (1-t)\mathbf{s}(0,0) + (t)\mathbf{s}(0,1), \\
\mathbf{R}_B(1,t) &= (1-t)\mathbf{s}(1,0) + (t)\mathbf{s}(1,1),
\end{aligned}$$

and are curves of the surface $\mathbf{P}_c(s,t)$ which should be subtracted from $(\mathbf{P}_a(s,t) + \mathbf{P}_b(s,t))$.

Further, we should define that a *degenerate Coons patch* is a patch of which one of the curves of a boundary has degenerated to a point. Instead of a curve, a point is used for one of the curves of (7.4).

7.6.2. Problems with Coons Method in Two-Dimensional Space.

The results with the linear Coons interpolation method in the two-dimensional space can be very unsatisfactory. In fact, the Coons interpolation, if used in two-dimensional space, is a projection of the Coons interpolation from three-dimensional space. When this projection takes place under the wrong angle, problems will occur. A closer inspection of these problems reveals that they occur when the linear interpolation of the surfaces $\mathbf{P}_a$ and $\mathbf{P}_b$ goes over the excess boundary curves $\mathbf{R}_A(s,0)$, $\mathbf{R}_A(s,1)$, or $\mathbf{R}_B(0,t)$, $\mathbf{R}_B(1,t)$, respectively, of surface $\mathbf{P}_c$. If this is the case, the subtracting surface $\mathbf{P}_c(s,t)$ of (7.9) is not limited to the area defined by $\mathbf{P}_a(s,t)$ or $\mathbf{P}_b(s,t)$. Parts of these surfaces will then lie *outside* the excess curves. This is the case in the following situations.

1. If the curves have one or more inflection point(s) (see Fig. 7.3).

2. If the integrated curvature of (one of) the curves is greater than 180° (see Fig. 7.4).

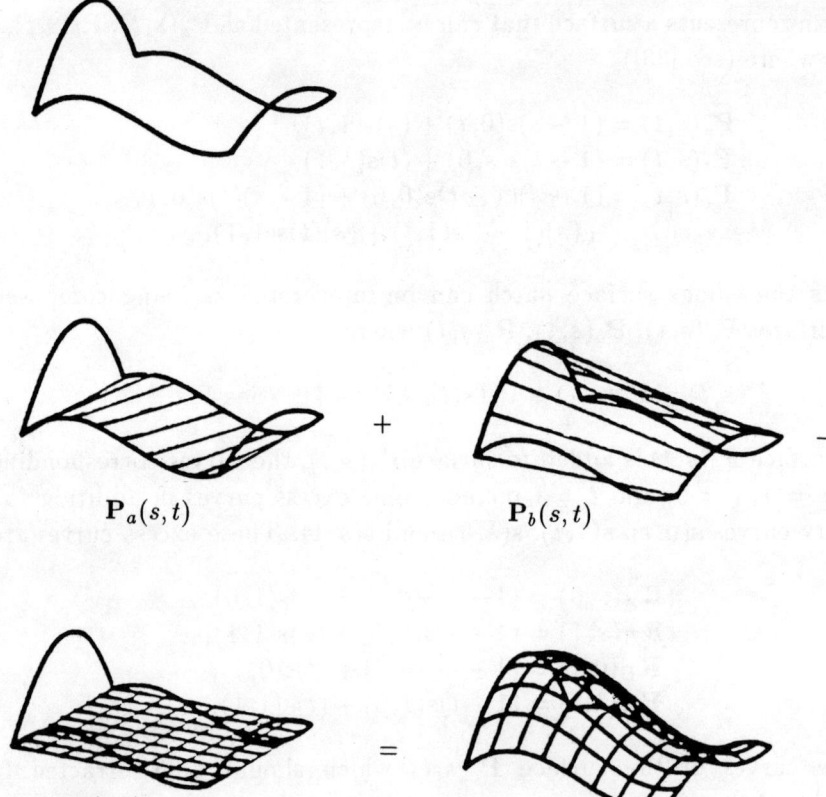

FIG. 7.2. *Composition of a Coons surface patch* [24, p. 139].

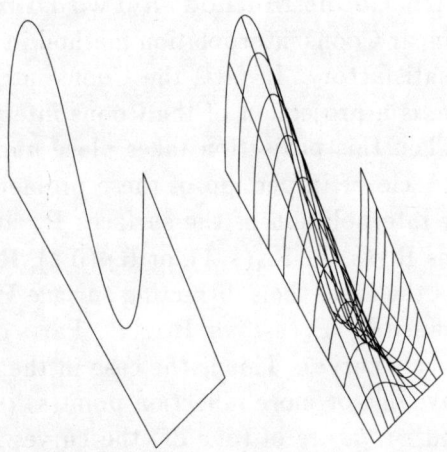

FIG. 7.3. *Problems with the two-dimensional linear Coons interpolation of four boundary curves when one curve has an inflection point.*

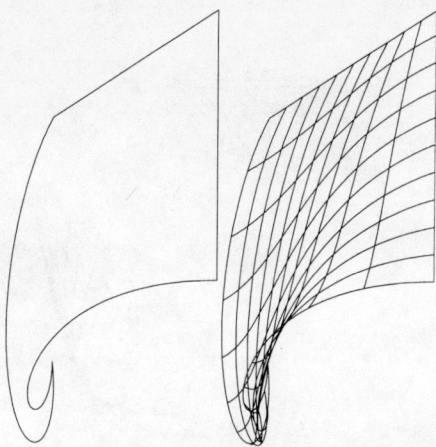

FIG. 7.4. *Problems with the two-dimensional Coons interpolation of four curves of which one has an integrated curvature greater than 180°.*

3. If a concave[3] curve is present a difference can be made between the problems caused by concave curves of four-sided Coons patches and degenerated foursided (triangular) Coons patches:

    a. All interpolations *can* go wrong if one or more concave curves are used for four-sided patches (see Fig. 7.5).

For three-sided degenerated Coons patches problems *can* occur if

    b. All three curves are concave. This is difficult to see in a picture but we have found some examples of it in the Coons calculations.

    c. Two curves are concave and the degenerated curve with length zero is opposite to one of the concave curves (see Fig. 7.6).

    d. One curve is concave and the degenerated curve with length zero lies opposite to the concave curve (see Fig. 7.7).

4. If the angle at the joint of two successive curves is greater than 180° problems with the interpolation can occur (see Fig. 7.8).

5. If opposite curves do not have the same degree, the Coons interpolation cannot be executed.

As stated before, these problems occur because we use a three-dimensional interpolation method for a two-dimensional problem. We also tried to use a nonlinear Coons interpolation with blending functions of degree 3 but the

---

[3]We will define a *concave* nonrational Bézier curve to be a curve of which the control points all lie to the left of the direction vector of the connection line between the begin and end control point of the curve provided that the control points of the Bézier curve are stored counterclockwise and no inflection points are available (see Fig. 7.12). In a similar way a *convex* nonrational Bézier curve is a curve of which all control points lie to the right of the direction vector of the connection line between the begin and end point provided that control points are stored counterclockwise and no inflection points are available.

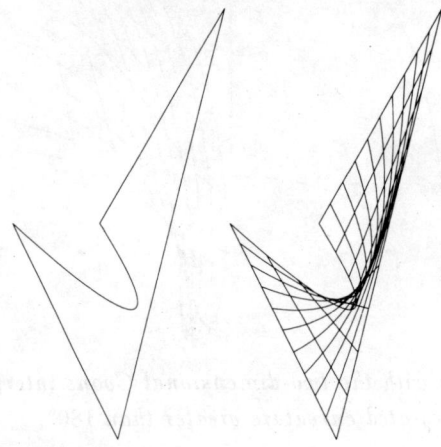

FIG. 7.5. *Problems with the two-dimensional linear Coons interpolation if one curve is concave.*

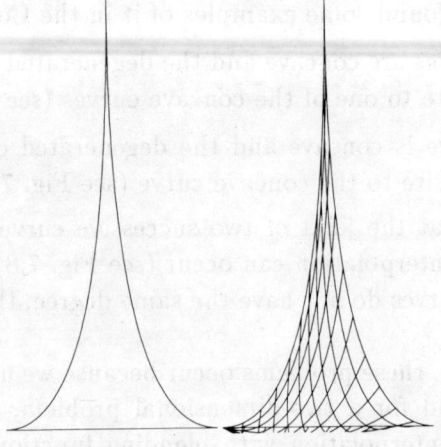

FIG. 7.6. *Problems with the two-dimensional linear Coons interpolation if two curves are concave.*

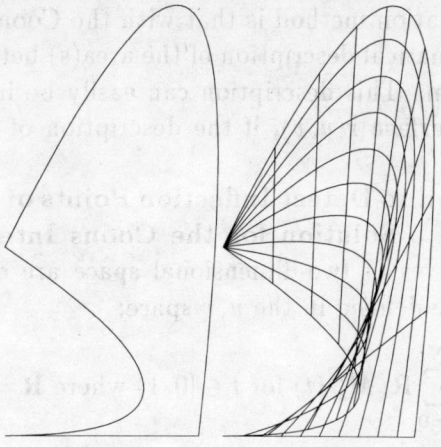

FIG. 7.7. *Problems with the two-dimensional linear degenerated Coons interpolation if one curve is concave.*

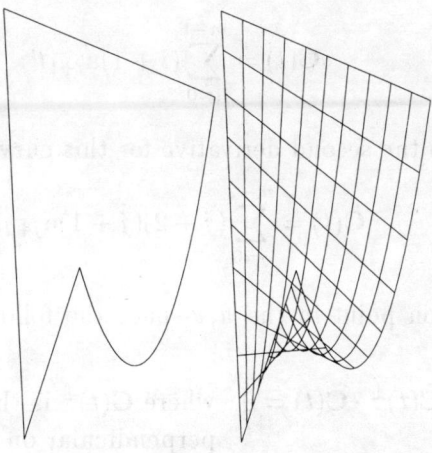

FIG. 7.8. *Problems with the two-dimensional linear Coons interpolation of four curves of which two successive curves have an angle greater than 180°.*

problems became even greater and more unpredictable. It would be a better idea to use a two-dimensional interpolation method. The reason to use the bilinear Coons interpolation instead of looking for a completely different two-dimensional interpolation method is that with the Coons interpolation we are able to get a mathematical description of the area(s) between boundary curves in a polynomial form. This description can easily be inserted in the original description of the surface $\mathbf{r}(u,v)$, if the description of this surface is also in polynomial form.

7.6.2.1. Method to Detect Inflection Points of a Two-Dimensional Bézier Curve and a Solution for the Coons Interpolation. Inflection points of Bézier curves in two-dimensional space are determined as follows. Given a Bézier curve defined in the u,v-space:

$$(7.11) \qquad \mathbf{C}(t) = \sum_{k=0}^{N} \mathbf{R}_k M_k^N(t) \text{ for } t \in [0,1) \text{ where } \mathbf{R} = \left(\begin{array}{c} R_u \\ R_v \end{array} \right).$$

To find the inflection points of this curve the next steps should be followed.

1. Convert the curve into the natural or monomial polynomial representation (see [12]) to get a curve with the same geometry defined over the same interval $t \in [0,1)$:

$$(7.12) \qquad \mathbf{C}(t) = \sum_{i=0}^{m} \mathbf{a}_i t^i \text{ where } \mathbf{a}_i = \left(\begin{array}{c} a_u \\ a_v \end{array} \right).$$

2. Determine the first derivative of this curve:

$$(7.13) \qquad \dot{\mathbf{C}}(t) = \sum_{i=0}^{m-1} (i+1)\mathbf{a}_{i+1} t^i.$$

3. Determine also the second derivative for this curve:

$$(7.14) \qquad \ddot{\mathbf{C}}(t) = \sum_{j=0}^{m-2} (j+2)(j+1)\mathbf{a}_{j+2} t^j.$$

4. For an inflection point in the u,v-space the following condition should be fulfilled:

$$(7.15) \qquad \dot{\mathbf{C}}(t)^{\perp} \cdot \ddot{\mathbf{C}}(t) = 0 \quad \text{where } \dot{\mathbf{C}}(t)^{\perp} \text{ is the vector}$$
$$\text{perpendicular on } \dot{\mathbf{C}}(t).$$

Because there are two solutions possible for this perpendicular vector, we choose the following one:

$$(7.16) \qquad \left(\begin{array}{c} A_{ui} \\ A_{vi} \end{array} \right) = \left(\begin{array}{c} -a_{vi} \\ a_{ui} \end{array} \right) = \mathbf{A}.$$

Thus, $\dot{\mathbf{C}}(t)^{\perp} = \sum_{i=0}^{m-1} (i+1)\mathbf{A}_{i+1} t^i.$

5. Substitute (7.14) and (7.16) into (7.15) as follows:

$$\sum_{i=0}^{m-1}(i+1)\mathbf{A}_{i+1}t^i \cdot \sum_{j=0}^{m-2}(j+2)(j+1)\mathbf{a}_{j+2}t^j = 0,$$

which can be written as

$$(7.17) \qquad \sum_{i=0}^{m-1}\sum_{j=0}^{m-2} t^{i+j}(i+1)(j+2)(j+1)(\mathbf{A}_{i+1}\cdot\mathbf{a}_{j+2}) = 0$$

from which the values of t can be determined. The algorithm of Boehm is used to divide the Bézier boundary curves into more parts. This is done at the detected inflection point(s). For higher degrees we found this solution is very unstable. Therefore, we used a different method. This method is derived from the field of computational geometry.

Given the Bézier control point polygon which is not self-intersecting, for all the control points the vector product of $\mathbf{P}_i\mathbf{P}_{i+1}$ with $\mathbf{P}_i\mathbf{P}_{i+2}$ is determined. In two-dimensional space the vector product can be used to determine the rotation of two vectors with respect to each other.

$$(7.18)\qquad
\begin{aligned}
L = \mathbf{a} \times \mathbf{b} &= |\mathbf{a}|\,|\mathbf{b}|\sin\alpha \\
\text{if } L > 0 &: \text{ the rotation is to the left,} \\
L < 0 &: \text{ the rotation is to the right,} \\
L = 0 &: \text{ the vectors are parallel.}
\end{aligned}$$

If L goes from greater than 0 to less than 0 or vice versa, an inflection point is possibly available (see Fig. 7.9). The Bézier curve is then divided into two curves at $t = 1/2$ with the Boehm algorithm and again for each curve; the possibility of an inflection point is checked. A disadvantage of this method is that a curve can be divided unnecessarily.

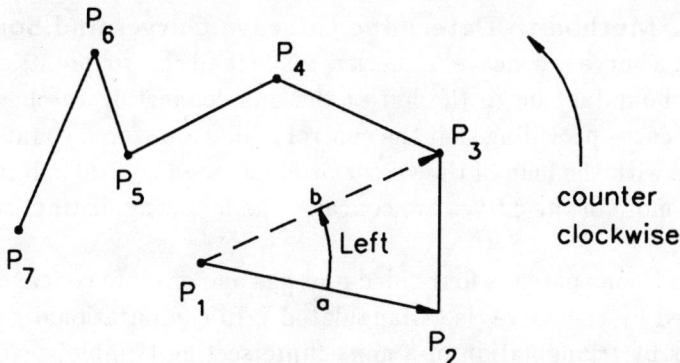

FIG. 7.9. *Determination of possible inflection points.*

7.6.2.2. Method to Detect the Integrated Curvature of Bézier Curves and a Solution for the Coons Interpolation. The integrated curvature of a Bézier curve of a boundary is determined with the help of the scalar (or dot) product used on the control point polygon of the Bézier curve. The first vector of the control point polygon is compared with the vectors of the rest of the polygon. With the scalar product the angle between the direction vectors of two lines can be determined.

(7.19)
$$S = \mathbf{a} \cdot \mathbf{b} = a_u b_u + a_v b_v = |\mathbf{a}|\,|\mathbf{b}| \cos \alpha$$
$$\text{if } S > 0 \text{ then } 0 \leq \alpha < 90°,$$
$$S < 0 \text{ then } 90° < \alpha \leq 180°,$$
$$S = 0 \ \alpha = 90°.$$

If the integrated curvature of the curve of a boundary is greater than 180° the curve is divided into two curves at $t = 1/2$. Again for each new curve a check is made if the rotation is less than 180°. If this is not the case, the curve is again divided into two curves and the procedure is repeated (see Fig. 7.10).

This means that only concave and/or convex curves with a rotation less than 180° will be available in the end.

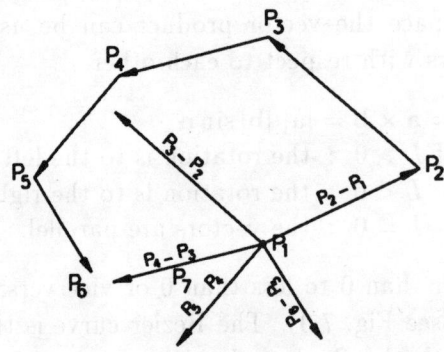

FIG. 7.10. *Determination of the integrated curvature of a two-dimensional Bézier curve.*

7.6.2.3. Method to Determine Concave Curves and Solutions. To determine if a curve is concave, a check is made if all control points of the Bézier curves of a boundary lie to the left of the line connecting the begin and end point of the curve providing that the control points are stored counterclockwise. This is done with the help of the vector product (see Fig. 7.11). If it is detected that one or more of the curves are concave, the following distinct solutions are offered.

1. If the Coons patch is four-sided and has one or more concave curves, the area enclosed by the curves is "triangulated." In computational geometry one understands by triangulation of a nonselfintersecting (simple) *polygon P* with n vertices [30]: "The finding of a set of $n - 3$ nonintersecting line segments joining nonadjacent vertices of P such that (a) the line segments lie within P, and (b) $n - 2$ nonoverlapping triangles are formed."

The "triangulation" of four curves can now be defined as follows. "Given four successive connected, nonselfintersecting, nonrational Bézier boundary

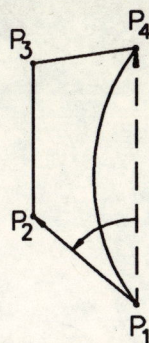

FIG. 7.11. *Determination of a concave curve.*

curves $c_i(i = 0, \cdots, 3)$ of which the first control point of the first curve is connected with the end control point of the last curve, find one nonintersecting line segment that joins nonadjacent begin and end control points of c_i in such a way that (a) the line segment lies within c_i, and (b) two nonoverlapping triangular area patches are formed."

The triangulation method goes now as follows (see Fig. 7.13).

a. Take the first two curves and check if the following conditions are fulfilled.

    i. Check if the two successive Bézier boundary curves have a convex angle $\alpha(< 180°)$ at their joint (see Fig. 7.12).

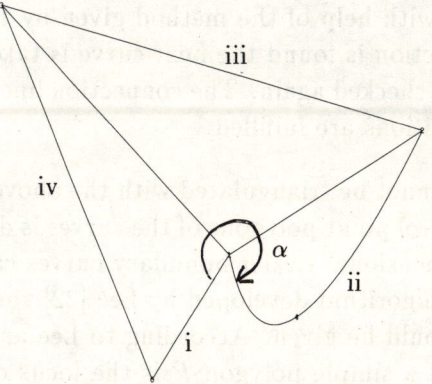

FIG. 7.12. *Reflex internal angle.*

    If the angle is not convex the next curve is taken and the procedure repeated.

    ii. Check if a control point of the two control point polygons lies on the left or on the connection line P_1P_2 of the begin point of the first curve and the end point of the second curve providing that the control points of the curves are stored counterclockwise and that the curves are only convex or concave.

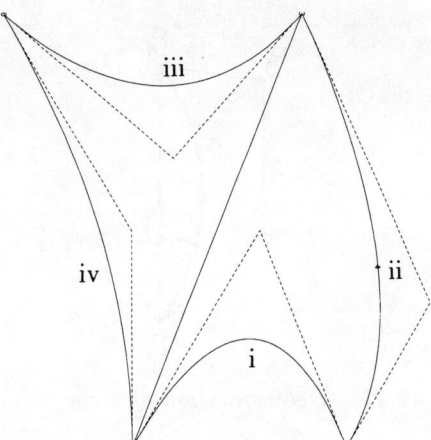

FIG. 7.13. *"Triangulation" of curves.*

This can be done with the help of the vector product between the direction vector of the connection line $\mathbf{P_1P_2}$ and the direction vectors of the other control points. If $L = \mathbf{a} \times \mathbf{b} \geq 0$, a point lies to the left or on the connection line and the connection line is rejected. The next curve is taken and the conditions i and ii are checked again.

  iii. If no control point lies to the left of this connection line, check if the connection line intersects with the other of the control point polygons with help of the method given by Pavlidis [26, p. 329]. If an intersection is found the next curve is taken and conditions i, ii, and iii are checked again. The connection line $\mathbf{P_1P_2}$ can be "drawn" if all conditions are fulfilled.

 b. If the curves cannot be triangulated with the above procedure, the kernel of the four control point polygons of the curves is determined. The kernel of four two-dimensional Bézier boundary curves can be determined with the help of an algorithm developed by Lee [22] and Preparata [27]. First a definition should be given. According to Lee and Preparata [22], "the kernel $K(P)$ of a simple polygon P is the locus of points internal to P which can be joined to every vertex of P by a segment totally contained in P. Equivalently, if one considers the boundary of P as a counterclockwise directed cycle, the kernel of P is the intersection of all halfplanes."

The kernel of the four Bézier boundary curves is now determined as follows (see Fig. 7.14).

  i. Check which curve is convex and which one is concave.

  ii. The control point polygon of convex curves is temporarily replaced by the connection line between the begin and end control point under the condition that the connection line does not intersect

with the other control point polygons. Otherwise the control
point polygon is normally used. This replacement is done because
otherwise the kernel may lie (partly) outside the boundary curves.
The control point polygon of concave curves is normally used.

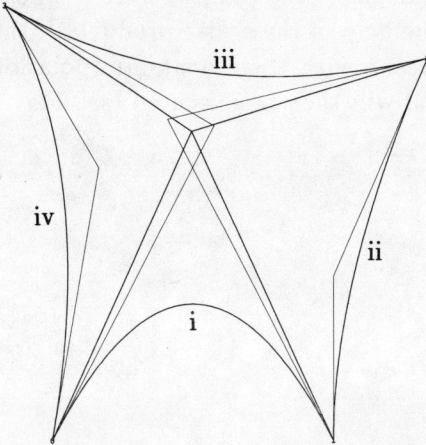

FIG. 7.14. *Determination of the kernel of four two-dimensional Bézier boundary curves.*

  iii. Now the kernel is determined on the polygon formed by the control
       points of the concave curves and the connection line of the convex
       curves.

Because the kernel is always convex, the average of its cornerpoints
results in a point within the kernel. This point is connected with
the 4 cornerpoints of the Coons patch. This subdivision results in 4
degenerated Coons patches.

c. If it is impossible to determine the kernel, the concave curves are one
   by one divided at $t = 1/2$ into two parts and a new "triangulation"
   procedure is started.

If the kernel can not be determined, split the curve with the longest
connectionline between the begin and end control point at $t = 1/2$. Start
the triangulation as described above, again.

2. If the Coons patch is degenerate and one or two boundary curves are
concave, the degenerated curve or control point may not lie opposite to these
concave curves.

3. If the Coons patch is degenerated and the three curves are concave, the
kernel of the three control point polygons is determined. Again, an average
point within the kernel can be determined by taking the average of the kernel
corner points. This average point is connected with the three corners of the
degenerated Coons patch. The result is three degenerated Coons patches each
having one concave curve.

4. If a reflex vertex is detected at the join of two successive curves, it will influence the "triangulation" or "quadrangulation" method. The latter method will be discussed in the coming paragraph.

7.6.2.4. Method to Determine the Internal Angles between Two Successive Curves. The internal angle between two successive curves can be determined by the angle of the tangent lines at their joint. This angle can be determined with the help of the scalar product. If the angle is greater than 180°, problems will occur with the Coons interpolation. A triangulation as described under 7.6.2.3 will then be executed (see Fig. 7.15).

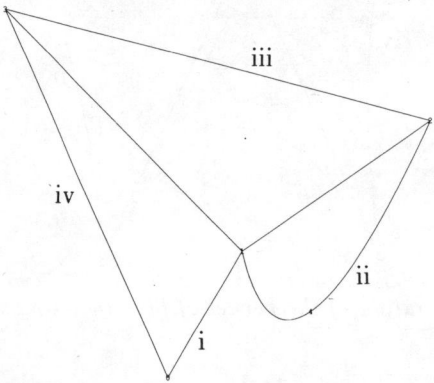

FIG. 7.15. *Triangulation of four curves, of which two successive curves have an internal angle greater than 180°.*

7.6.3. "Quadrangulation" of Curves. If more than four Bézier boundary curves are given, a "triangulation" of curves will result in quite a number of sets of curves and connection lines, which should be interpolated to find the definition of the area between these curves. Thus, quite a number of surfaces can be formed. So, we looked for a procedure that results in a smaller number of surfaces, namely sets of four and, if necessary, three curves that can be interpolated by a linear Coons interpolation. We called this method the "quadrangulation" of two-dimensional Bézier curves.

One of the aims of scientists working in the field of computational geometry and dealing with the triangulation of polygons is to make triangulation algorithms much faster by exploiting special characteristics of the polygons. Fast algorithms are available for, e.g., star-shaped polygons [30], [21], [15]. We were interested in a possibility of "quadrangulating" a set of nonselfintersecting Bézier boundary curves instead of triangulating polygons and, if possible, using a fast algorithm. Therefore, special characteristics of our two-dimensional Bézier curves should be used, if possible. We found that the Bézier boundary curves in our algorithm will have the following characteristics:

1. If curves have an integrated curvature that is too large or if they have inflection points, we decided to divide these curves into more curves in such a

way that in fact only concave or convex curves are left. The reasons for doing this are caused by the problems we had with the Coons interpolation used in two-dimensional space. Thus, all the curves are convex or concave and do not have a integrated curvature greater than 180° or inflection points.

2. The curves are not selfintersecting and do not intersect each other except at their joints.

3. The begin point of the first curve and the end point of the last curve coincide.

4. Looking at all control point polygons of all curves, no special characteristics (such as star-shaped, point visible, etc. [10]) can be exploited because they are unknown.

Therefore, the conclusion is that no fast existing triangulation algorithm for polygons should be modified to get a "quadrangulation" algorithm for curves. We modified an algorithm of Meisters [23]. The "quadrangulation" of curves can be defined as follows. "Given n successive connected, nonselfintersecting, nonrational Bézier boundary curves c_i of which the first control point of the first curve is connected with the end control point of the last curve, find a set of $n - 4$ nonintersecting line segments which join nonadjacent begin and end control points of c_i in such a way that (a) the line segments lie within c_i, and (b) nonoverlapping quadrangular and triangular area patches are formed."

If the control point polygons of successive Bézier boundary curves are stored counterclockwise, the algorithm works as follows (see Fig. 7.16).

1. Determine the total number of curves.

2. Take the first three curves $c_i(t)$ to $c_{i+2}(t)$.

3. Check if the internal angles at the two joints of the three curves are less than or equal to 180°. If this is not the case, skip one curve and take curves $c_{i+1}(t)$ to $c_{i+3}(t)$.

4. Determine the line $P_1 P_2$ connecting the first control point of the first curve $c_i(t)$ and the last control point of the third curve $c_{i+2}(t)$.

5. Check if the control point polygons of the three or four curves intersect with the connection line $P_1 P_2$. This can be done with the intersection of line segments method described by Pavlidis [27]. If this is the case, skip one curve and take curves $c_{i+1}(t)$ to $c_{i+3}(t)$ and start the procedure again.

6. Check if the connection line $P_1 P_2$ intersects with the rest of the control point polygons of all other boundary curves. If this is the case, reject $P_1 P_2$ skip one curve, and take curves $c_{i+1}(t)$ to $c_{i+3}(t)$ and start the procedure again at point 2. Otherwise the connection line $P_1 P_2$ can be "drawn."

7. If a resulting quadrangle consists of four curves, check if one or more curves of a quadrangle are concave. If so, triangulate the quadrangle and go to 8.

If the quadrangle cannot be triangulated,

- Determine the kernel,

- Determine a point P in the kernel,

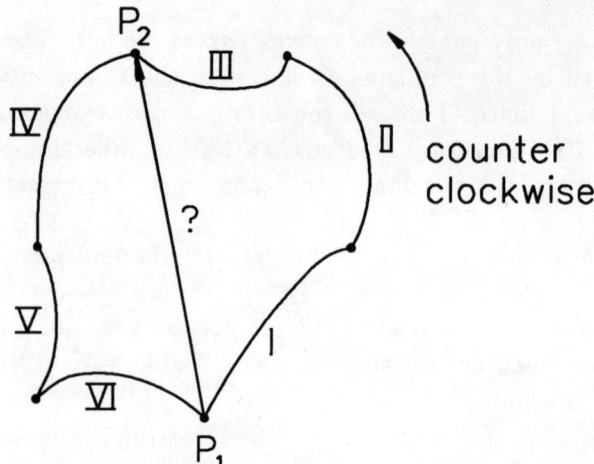

FIG. 7.16.   *Quadrangulation of nonrational Bézier curves.*

- Connect this point **P** with the corner points of the quadrangle,

- Go to 8.
If the kernel cannot be determined,
- Determine the length of the connection lines between the begin control
  point and the end control point of each control point polygon,

- Split the curve with the longest connection line at $t = 1/2$,

- Go to 2 to start the quadrangulation again.
8. If the resulting quadrangle consists of three curves, check if all curves
are convex. If this is the case, take the next three curves c_{i+3} to c_{i+5} and go
to 2.
If one curve is concave,
- Choose the degenerated curve to be opposite of the concave curve,

- Take the next three curves c_{i+3} to c_{i+5} and go to 2.
If two curves are concave,
- Choose the degenerated curve with length zero to be opposite of the
  nonconcave curve or straight line,

- Take the next three curves c_{i+3} to c_{i+5} and go to 2.
If three curves are concave,
- Determine the kernel of the control point polygons,

- Determine a point **P** in the kernel,

- Connect **P** with the corner points of the triangular patch. This results
  in three triangular patches each having one concave curve. Go to 8.

7.7. The Algorithm

The final algorithm consists now of the following steps.

1. Store the control points of successive boundary Bézier curves counterclockwise.

2. Determine the inflection points per Bézier boundary curve and split each curve at these points.

3. Check the integrated curvature of each curve. If this is greater than 180° split the curve and check the integrated curvature again.

4. Check if the control point polygons intersect. If so, determine the length of the connection line between the begin and end point of each of these intersecting polygons. Split the curve with the longest connection line and repeat procedure 4.

5. Determine if the number of curves is less than three. If so, split the (one of) the curves until three or four curves are available.

6. "Quadrangulate" the curves.

7. Determine if the opposite boundary curves in each triangular or quadrangular patch have the same degree. If not, determine the highest degree M of two opposite curves. Elevate the degree of the other curve to this degree M.

8. Convert the Bézier basic surface, the Bézier boundary curves, and all connection lines into the polynomial representation.

9. Do the Coons interpolation as follows:

   a. Define two opposite boundary curves as $\mathbf{r}(s,0)$ and $\mathbf{r}(s,1)$ and the two other opposite boundary curves as $\mathbf{r}(0,t)$ and $\mathbf{r}(1,t)$ (see also (7.4) and (7.10)):

$$
\mathbf{r}(s,0) = \mathbf{c}_k(w) = \sum_{i=0}^{MS} \mathbf{a}_i s^i \quad \mathbf{r}(0,t) = \mathbf{c}_{k+1}(w) = \sum_{j=0}^{MT} \mathbf{d}_j t^j,
$$

(7.20)

$$
\mathbf{r}(s,1) = \mathbf{c}_{k+2}(w) = \sum_{i=0}^{MS} \mathbf{a}_i s^i \quad \mathbf{r}(1,t) = \mathbf{c}_{k+3}(w) = \sum_{j=0}^{MT} \mathbf{d}_j t^j.
$$

   b. Now the bilinear Coons interpolation should be executed to find all the coefficients $\mathbf{b}_{f,g}$ (see (7.5)).

10. Substitute the resulting mathematical description of the Coons interpolation into the description of the basic surface (see (7.6)).

11. Determine the new coefficients of each "resulting surface."

12. Convert the "resulting surfaces" back into the Bézier representation.

7.8. Examples

Below a few examples of this algorithm are given.

1. The boundary of an arbitrary figure is given in the u,v-space and the area between the edges of the u,v-space and these boundary curves is

determined with the help of the algorithm. The resulting surfaces are inserted in the basic surface $\mathbf{r}(u, v)$. See Fig. 7.17.

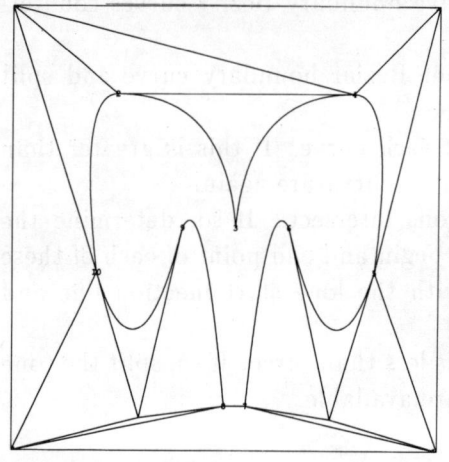

FIG. 7.17. *Algorithm used to determine the area between the boundary curves and the edges of the u, v-space. The result is inserted in the basic surface.*

2. In this example the area between the boundary curves is determined with the help of the algorithm. The result is again inserted in the same basic surface with the following result (see Fig. 7.18).

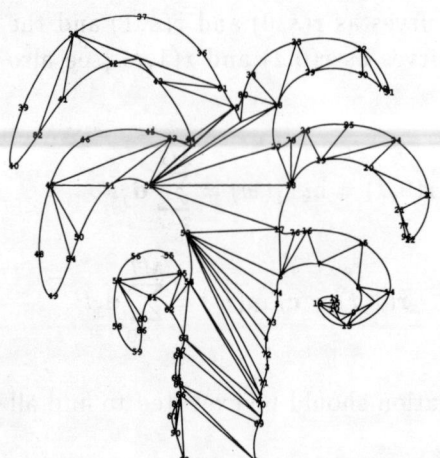

FIG. 7.18. *Algorithm used to determine the area between the boundary curves. The result is inserted in the basic surface.*

7.9. Unsolved Problems

The algorithm fulfills the conditions and requirements stated in §7.5. We implemented the algorithm and it works successfully but we still have a number of problems that need to be solved.

1. The degree of the resulting surfaces can become very high. If the basic surface is of degree 3 and the boundary curves in the u, v-space are also of degree 3, the resulting surface(s) can have a maximum degree of 18 in both directions. So, for higher degrees of the basic surface and/or the boundary curves, the degree of the "resulting surfaces" will become greater than 20, which is unacceptable for most CAD systems. Only a degree reduction (see [17]) of the basic surface and/or of the boundary curves before our algorithm is used can solve this problem. The basic surface will consist after this degree reduction of a composite Bézier surface with a regular rectangular topology that is trimmed by boundary curves. We did not solve the problem of trimmed composite Bézier surfaces yet.

2. A high degree also causes problems with intersection calculations, which can become very unstable (see [12]).

3. If a great number of surfaces is the result, modeling with the surfaces can cause problems.

7.10. Conclusions

An algorithm was presented to find the explicit representation of the surface between nonrational Bézier boundary curves that trim a nonrational Bézier basic surface. The algorithm was implemented and works satisfactorily. The algorithm fulfills the conditions stated in §7.5. We think that the problems with the high degree of the "resulting surfaces" can be solved by a degree reduction of the boundary curves and the basic surface before our algorithm is used.

Quite a number of objections can be formulated against this algorithm, such as high degree of the "resulting surfaces," the total number of surfaces, the irregular topology, etc. We should keep in mind that this algorithm has been developed to make data exchange possible between dissimilar CAD systems, one having the function of trimmed entities and the other not having this function. Further, NC machines in general do not recognize trimmed surfaces. We think that in these cases our solution is useful. As soon as CAD systems and NC machines are more advanced, this method loses its value. Another approach would have been to approximate the surface enclosed by the boundary. The questions are, if with such an approach the total number of "resulting surfaces" would be smaller; if the degree of these surfaces would be lower and; if a rectangular topology is used, if the deviation of the "resulting surfaces" with the original surface is smaller than when an irregular topology is used in combination with an exact conversion.

Acknowledgments

We would like to thank Prof. Dr. Ing. H. Nowacki, Prof. D. J. McConnalogue, Dr. J. S. M. Vergeest, and Ir. E. Boender for their critical remarks, which were of great help to improve this paper. This research is supported by the Netherlands Technology Foundation (STW).

References

[1] W. Boehm, *Inserting new knots into B-spline curves*, Comput. Aided Des., 12 (1980), pp. 199–201.

[2] ——, *Generating Bézier points of B-spline curves and surfaces*, Comput. Aided Des., 13 (1981), pp. 365–366.

[3] ——, *On efficiency of knot insertion algorithms*, Comput. Aided Geom. Des., 2 (1985), pp. 141–143.

[4] W. Boehm and H. Prautzsch, *The insertion algorithm*, Comput. Aided Des., 17 (1985), pp. 58–59.

[5] M. S. Casale, *Free-form solid modeling with trimmed surface patches*, IEEE Comput. Graphics Appl. (1987), pp. 33–43.

[6] M. S. Casale and J. E. Borrow, *A set operation algorithm for sculptured solids modeled with trimmed patches*, Comput. Aided Geom. Des., 6 (1989), pp. 237–247.

[7] E. Cohen and R. F. Riesenfeld, *General matrix representations for Bézier and B-spline curves*, Computers in Industry, 3 (1982), pp. 9–15.

[8] S. A. Coons, *Surfaces for computer aided design of figures*, M.I.T. ESL 9442-M-139, 1964.

[9] ——, *Surfaces for computer aided design of space forms*, Report MAC-TR-41, MIT,1967.

[10] H. El Gindy and G. T. Toussaint, *On geodesic properties of polygons relevant to linear time triangulation*, The Visual Computer, 5 (1989), pp. 68–74.

[11] G. Farin, *Curves and Surfaces for Computer Aided Geometric Design: A Practical Guide*, Academic Press, Boston, 1988.

[12] R. T. Farouki, *Trimmed surface algorithms for the evaluation and interrogation of solid boundary representations*, IBM J. Res. Develop., 31 (1987), pp. 314–334.

[13] R. T. Farouki and V. T. Rajan, *On the numerical condition of polynomials in Bernstein form*, Comput. Aided Geom. Des., 4 (1987), pp. 191–216.

[14] ——, *Algorithms for polynomials in Bernstein form*, Comput. Aided Geom. Des., 5 (1988), pp. 1–26.

[15] W. C. Ho, *Decomposition of a polygon into triangles*, The Mathematical Gazette, 59 (1975), pp. 132–134.

[16] M. Hosaka and F. Kimura, *A theory and methods for free form shape construction*, J. Inform. Process., 3 (1980), pp. 140–151.

[17] J. Hoschek and F. J. Schneider, *Spline conversion for trimmed rational Bézier and B-spline surfaces*, Comput. Aided Des., 22 (1990), pp. 580–590.

[18] ISO, *Industrial automation systems exchange of product model data representation and format description*, First working draft, ISO TC 184/SC4/WG1, 1988.

[19] K. J. Kelder, *Evaluation of a trimmed surface*, Master's thesis, Faculty of Industrial Design Engineering, Technical University Delft, the Netherlands, 1988, 124 pp., in Dutch.

[20] K. J. Kelder, A. A. M. Ranke, and J. S. M. Vergeest, *Transmission of trimmed surfaces between incompatible CAD systems*, in Ontwikkelingen rond industriële automatisering, Congres bijdragen CAPE '89, Samson Uitgeverij, Alphen aan den Rijn, Brussel, 1989, pp. 201–215, in Dutch.

[21] S. H. Lee and K. Y. Chwa, *A new triangulation linear class of simple polygons*, Internat. J. Comput. Math., 22 (1987), pp. 135–147.

[22] D. T. Lee and F. P. Preparata, *An optimal algorithm for finding the kernel of a polygon*, J. Appl. Comput. Math., 26 (1979), pp. 415–421.

[23] G. H. Meisters, *Polygons have ears*, Amer. Math. Monthly (1975), pp. 648–651.

[24] G. Mullineux, *CAD: Computational Concepts and Methods*, Kogan Page, London, 1986.

[25] J. Owen, *Introduction, scope and definitions*, ISO TC184/SC4/WG1 Nr. 283, 1988.

[26] T. Pavlidis, *Algorithms for Graphics and Image Processing*, Springer-Verlag, Berlin–Heidelberg, 1982.

[27] F. P. Preparata and M. I. Shamos, *Computational Geometry. An Introduction*, Springer-Verlag, New York, 1985.

[28] A. P. Rockwood, *In an age of surfaces*, Iris Universe (1989), pp. 12–22.

[29] A. P. Rockwood, T. Davis, and K. Heaton, *Real time rendering of trimmed surfaces*, Proceedings of Siggraph '89, Boston, 1989.

[30] A. Schoone and J. van Leeuwen, *Triangulating a star-shaped polygon*, Tech. Rep. No. RUV-CS-80-3, University of Utrecht, the Netherlands, Utrecht, 1980.

[31] J. J. Shah and P. R. Wilson, *Analysis of design abstraction, representation and inferencing requirements for computer aided design*, Design Studies, 10 (1989), pp. 169–178.

[32] T. C. Woo and S. Y. Shin, *A linear time algorithm for triangulating a point visible polygon*, ACM Trans. Graphics, 4 (1985), pp. 60–70.

[33] F. Yamaguchi, *Curves and Surfaces in Computer Aided Geometric Design*, Springer-Verlag, Berlin, 1988.

A Procedural Feature-Based Approach for Designing Functional Surfaces

James C. Cavendish and Samuel P. Marin

8.1. Introduction

While many surfaces such as automobile outer panels, ship hulls, and airfoils are characterized by their smooth, free-form shapes, a far larger class of functional surfaces are characterized by highly irregular, multi-featured shapes consisting of pockets, channels, ribs, etc. For example, a typical car contains well over 200 stamped sheet metal parts, but only about 5% of these are free-form outer panels. The rest are multi-featured, functional inner panels (see Fig. 8.1). A parts explosion for an airplane, refrigerator, snow blower, lawn mower, etc., would reveal more or less the same balance of surface types as that indicated in Fig. 8.1. Some of the differences between free-form and functional surfaces can be seen by comparing the parts shown in Fig. 8.2 with those in Fig. 8.3.

In contrast to the design of aesthetic, free-form surfaces, functional surface design can perhaps best be viewed as a process of assembling a collection of known component surfaces to form a single composite surface. This composite surface is comprised of a base or primary surface into which a number of pocket-like geometric features have been embedded. The known component surfaces form the tops or bottoms of the various pockets and must be assembled so that the entire surface satisfies certain functional objectives. For example, the simple contact lens cleaning case shown in Fig. 8.3 contains two circular pockets which function as receptacles for plastic vials that come with the lens cleaning kit. The large L-shaped pocket in this surface serves as a receptacle for a strip of foil-wrapped cleaning tablets while the two depressions embedded in the legs of the L evidently function to ease access to the foils by thumb and index finger. For this functional surface, component surfaces (i.e., pocket tops and bottoms) are pieces of simple planes or cylinders. The shape of each piece is determined by bounding curves which, in this case, are either circles or rounded polygons.

Design of functional surfaces usually begins with a *plan view* drawing of the surface. Figure 8.4, for example, illustrates both perspective and plan view

145

Outer Panels
Inner Panels

FIG. 8.1. *Sheet metal parts explosion for a car.*

FIG. 8.2. *Examples of smooth, free-form surfaces. Exterior surface of an automobile (top) and teapot model (bottom).*

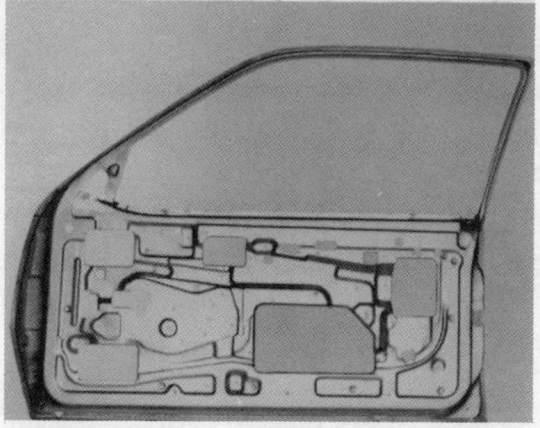

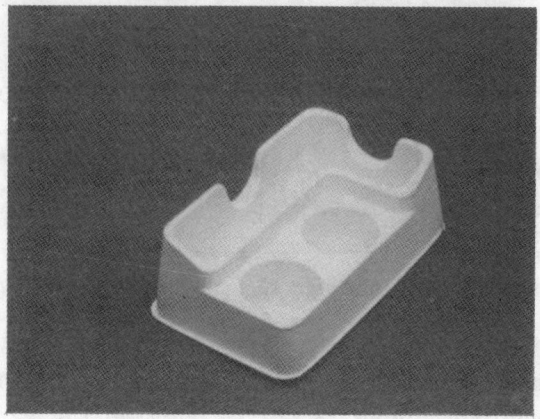

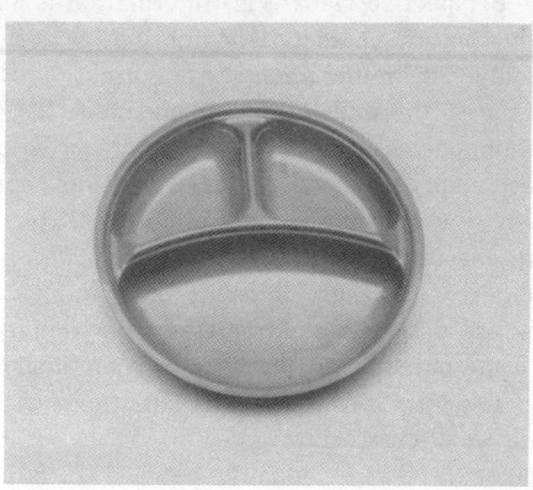

FIG. 8.3. *Examples of multi-featured, functional surfaces. Door inner panel (top), contact lens cleaning kit (center), and frozen food dish (bottom).*

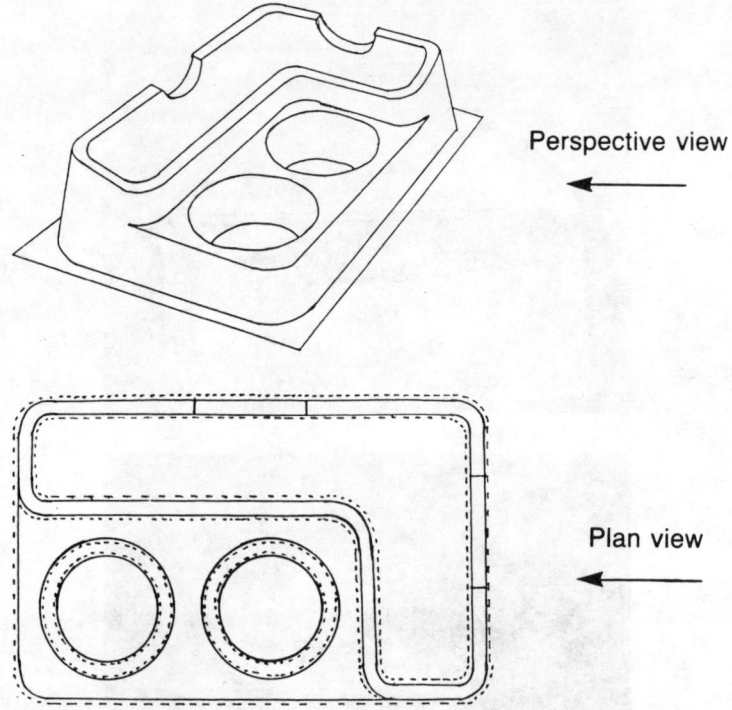

Perspective view

Plan view

FIG. 8.4.   *Perspective and plan view sketches of contact lens cleaning case.*

drawings of the contact lens cleaning case. The plan view drawing shows planar projections of the pocket boundary curves lying on the various component surfaces seen in the perspective view. This information is augmented with a small amount of other feature characteristic information such as fillet radii or wall angles. Using only this limited amount of information, the designer must create transition surfaces, the walls of the circular pockets, for example, that smoothly join the component surfaces to form the required pocket geometry. With the patch-based computer aided design (CAD) tools currently available, this can be extremely time consuming and quite difficult.

Indeed, current generation tools for computer aided geometric design (CAGD) do not address the special needs of functional surface design. Functional surfaces are designed and represented within commercial CAD systems using the same patch-based mathematical methods that are used to represent aesthetic, free-form surfaces. This situation has lead to four key problem areas which we discuss below.

First among these is that, generally, patched surfaces are inefficient representors of functional surfaces. Because of the vastly different length scales associated with most functional surfaces, patched representations require a large number of surface patches. While only a modest 10 or so patches would suffice to represent the door outer surface in Fig. 8.2 (32 patches for the teapot from [1], shown in the same figure), well over 1000 patches would be required to adequately capture the door inner surface geometry displayed in Fig. 8.3.

Next, patched surfaces are usually difficult to modify. Using patched-based representations, any design modification made to a functional surface entails rebuilding relevant patches. This means that the designer must redefine patch boundaries and adjust patch parameters to achieve a desired surface shape as well as surface smoothness.

Third, the continuity of derivatives across patch boundaries, required for NC (numerically controlled) machining applications, may be difficult to achieve in some CAD systems.

The forth and last point is that a patch-based data structure is poorly matched to the character of typical functional surfaces. The geometric building blocks used in patched-based modelers are parametric surface patches usually defined on rectangular grids in parameter space. Individual features on a functional surface can be represented only by assembling large numbers of these patches. Downstream processing of the surface (for example, NC path generation) must be done one patch at a time.

In this paper we describe a new *feature-based* approach to the design and representation of functional surfaces, tailored to overcome the four key shortcomings of patch-based techniques expressed above.

This feature-based approach provides a greatly simplified design and representation capability for functional surfaces. Surface feature details can be easily designed with a minimal number of input specifications which an engineer or designer naturally uses to describe the geometry of a particular functional surface. This approach also provides an easy modification capability along with accurate control of surface shape and a rigorous guarantee of surface smoothness. Morover, the resulting data structure is especially well suited to the character of the functional surfaces to be designed. The geometric building block in the proposed data structure is an entire *feature* rather than a single *patch*. This provides a more direct and natural approach to the modeling of functional surface characteristics.

In §8.2 we describe the essentials of the new approach. This discussion is accompanied by appropriate reference to current design practices and by several examples illustrating the design of prototype features and functional surfaces. §§8.3 and 8.4 are devoted to further development and extension of the basic approach outlined in §8.2. Additional examples of realistic functional surfaces designed on a prototype CAD system using the proposed approach are discussed in §8.5. Our conclusions are given in §8.6.

8.2. A New Approach to Functional Surface Design

Our feature-based approach turns on the simple observation, already stated, that many functional surfaces are formed, from a given base surface, by taking pieces of known surfaces (which we call secondary surfaces) and smoothly blending them, under a few additional specifications, to the base surface along given curves (the feature boundaries) to create the required features. Hence, our view of functional surface design starts with a primary or base surface which

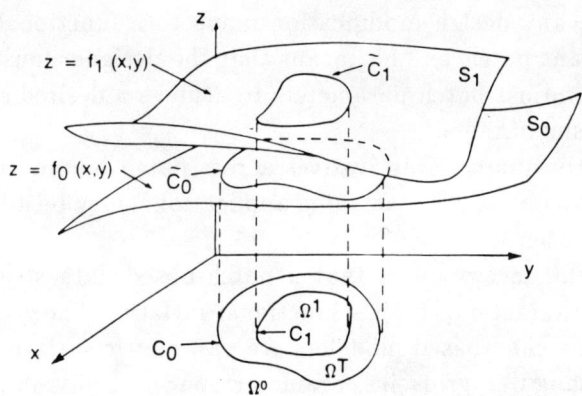

FIG. 8.5. *Outline of feature boundary curves on primary and secondary surfaces.*

is then modified by introducing a single pocket-like feature on the surface. If we can accomplish our goal once — build a single feature to specification on a base surface — we can do it again and again to design complex shapes by simply treating the modified surface as a new base surface to be again altered through the blending of a new component surface along specified boundaries. Thus, we will begin our exposition with an emphasis on single feature design.

8.2.1. Single Feature Design.

The starting point for the design of a single feature (pocket) is shown in Fig. 8.5.

The coordinate system has been selected so that the xy-plane corresponds to the plan view mentioned above. Most often, a single feature involves a specified *secondary* surface, S_1, the floor of a pocket, for example, which must be joined to a known *primary* surface, S_0, with a smooth transition occurring between two closed curves. One of these two curves, $\tilde{C}_0$, is specified on S_0 to start the transition and the second curve, $\tilde{C}_1$, is given on S_1 to complete the transition (see Fig. 8.5). Since we are concerned mainly with the design of stamped sheet metal parts, we assume that both S_0 and S_1 are representable in explicit form; that is, S_0 and S_1 can be described by functions $z = f_0(x, y)$ and $z = f_1(x, y)$, respectively. Of course, S_0 or S_1 may be given parametrically as long as they admit an explicit representation with respect to the plan view coordinates. We will discuss this issue further in §8.4.

We also assume, as shown in Fig. 8.5, that the plan view projections of $\tilde{C}_0$ and $\tilde{C}_1$, denoted here by C_0 and C_1, are simple closed curves that partition the xy-plane into three disjoint regions: the region Ω^0 outside C_0, the region Ω^1 inside the curve C_1, and a region Ω^T between these two curves. We wish to design a new surface, P_1, which is equal to S_0 on Ω^0, equal to S_1 on Ω^1, and which makes a smooth transition between S_0 and S_1 in the region Ω^T. Our strategy is to construct first a function $z = \Phi(x, y)$ which we call a *transition function* and which satisfies

$$\Phi : R^2 \rightarrow [0,1]$$

(8.1) $\qquad \Phi(x,y) = \begin{cases} 0 & \text{if } (x,y) \in \Omega^0 \\ 1 & \text{if } (x,y) \in \Omega^1 \end{cases}$

$\qquad\qquad \nabla\Phi$ is continuous in R^2.

The transition function provides a prototype for the desired feature. The precise definition of Φ on Ω^T will be provided shortly. For now we note that, once $\Phi(x,y)$ has been defined, we can easily construct a candidate for the desired single feature surface, P_1, given by the explicit representation $z = g_1(x,y)$ where

(8.2) $\qquad\qquad g_1(x,y) = (1 - \Phi(x,y))f_0(x,y) + \Phi(x,y)f_1(x,y).$

This simple blending formula provides a very useful tool to quickly define and represent single pocket-like features when component surfaces can be explicitly represented. When applied in recursive fashion, as we discuss next, (8.2) forms the basis for a simple but powerful procedural approach to the design and representation of complex functional surfaces.

8.2.2. Multiple Feature Design. Generalization of the single feature design in (8.2) to a multi-feature design is straightforward. We proceed as follows if the intent is to design a geometry involving a base surface $S_0 : z = f_0(x,y)$ with N features whose tops or bottoms are defined by component surfaces $S_i : z = f_i(x,y), i = 1, 2, \cdots, N$. In order to integrate the ith feature into the design, we must define the plan view shape of the ith feature and construct Φ_i, the transition function for the ith feature. Once these are available we then define N intermediate surfaces, $P_i : z = g_i(x,y), i = 1, \cdots, N$ via the recurrence relation:

(8.3) $\qquad\qquad g_i(x,y) = (1 - \Phi_i(x,y))g_{i-1}(x,y) + \Phi_i(x,y)f_i(x,y)$

where $g_0(x,y) \equiv f_0(x,y)$. The desired N-featured surface is P_N represented by $z = g_N(x,y)$.

To complete the description of the basic method we must say how plan view feature shapes are defined and how the corresponding transition functions are constructed. This is accomplished in the following two subsections.

8.2.3. The Definition of Pocket Boundaries. To define plan view pocket boundaries, we begin by inputting a sequence of vertices to define the approximate shape of the inner boundary curve, C_1, for the ith feature. For example, if a pocket is to be a rectangular shape—with rounded corners— we input the (x,y)–coordinates of four vertices, $(x_i, y_i), i = 1, \cdots, 4$, to define a piecewise linear approximation to the intended inner curve (designers refer to this construct as "framing in the feature"). Then, to control the degree of rounding at each corner, we input four radii, $r_i, i = 1, \cdots, 4$. The sharp corners are automatically replaced by smoothly fitted circular arcs with designated

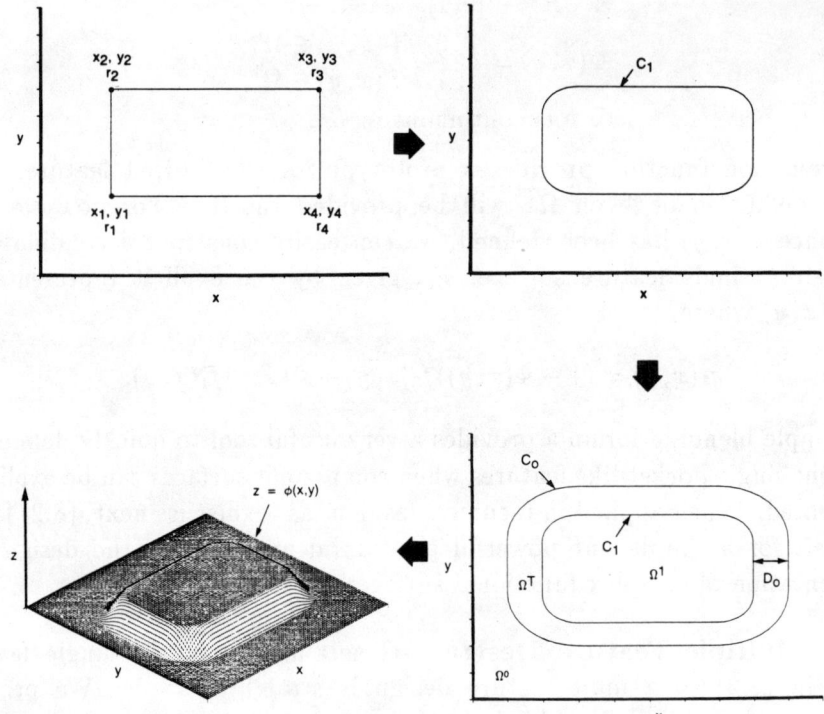

FIG. 8.6. *Construction of 4-sided feature boundary curves in plan view and associated transition function.*

radii of curvature. We obtain the outer curve, C_0, as a constant normal offset of the inner curve just constructed. (Variable normal offsets will be discussed in §8.3). This is done by displacing each point on the smoothed inner curve, C_1, by a given amount D_0, an input offset distance, in the direction of the unit outward normal to C_1. This construct for defining feature boundary curves in plan view is useful for both convex shapes and shapes with reentrant corners, provided the offset, D_0, is small enough. Figure 8.6 illustrates these geometric constructions.

8.2.4. The Definition of $\Phi(x, y)$. The pocket boundary construction given in §8.2.3 induces a natural local coordinate system in Ω^T which is used to provide a particularly simple definition of Φ. If the inner curve C_1 is described parametrically by $(x(s), y(s))$, $0 \leq s \leq L$, then to each point $(x, y) \in \Omega^T$ we associate coordinates (ρ, s) where ρ is the distance from (x, y) to C_1 and s is the arclength parameter associated with the orthogonal projection of (x, y) on C_1. A family of transition functions can now be defined by introducing $z = h(\rho)$ where $h(\rho)$ is a simple univariate function; for example, a cubic Hermite polynomial satisfying end point conditions $h(0) = 1, h'(0) = h(D_0) = h'(D_0) = 0$. For each fixed value of s, the slice of the surface $S : x = x(\rho, s), y = y(\rho, s), z = h(\rho)$ will then be a plane cubic curve. With this, $\Phi(x,y)$ assumes the definition $\Phi(x,y) = h(\rho(x,y))$ for $(x,y) \in \Omega^T$ in (8.2). Figure 8.6 displays a rectangular shaped transition function which blends together the surfaces $f_1(x,y) = 1$ and $f_0(x,y) = 0$ in (8.2).

8.2.5. Example. In this subsection we illustrate the application of the simple ideas presented so far to obtain a seven-step design sequence for the contact lens cleaning case displayed in Fig. 8.3. This sequence is shown in Fig. 8.7(a–g). All engineering design data needed to define the feature boundary curves (vertices, radii, offsets) were entered interactively at the terminal of a Silicon Graphics Personal Iris workstation on which the procedural algorithm defined by (8.3) was implemented in FORTRAN. Each of the seven surface iterates in the recursion was rendered in color by sampling it on a 150 x 150 rectangular grid, triangulating the surface on this grid, and executing graphics display routines developed by Dr. Paul Besl at the General Motors Research Labs. The complete design and rendition effort required less than half an hour at the terminal. Computation time required to evaluate the final seven-featured surface at all 22,500 points defined by the rectangular grid was about 20 seconds.

We conclude this section by indicating how surface feature details can easily be changed in order of affect a hypothetical design modification that may be required in a previously completed surface.

A disinfecting solution, dispensed from a small plastic bottle, is ordinarily used to clean contact lenses. Suppose now that a design change is called for in the surface displayed in Fig. 8.7(g) which will produce a modified surface that will contain an additional depression to function as a receptacle for the plastic bottle. One possible way to carry this out is by first lengthening the pocket displayed in Fig. 8.7(a). This requires only a simple translation of two of the vertices used to define Ω_1^T in the first surface iterate which in turn modifies Φ_1 in (8.3). A reevaluation of the complete procedural algorithm which includes only this small modification then produces the surface rendered in Fig. 8.8(a). This design change required only seconds to implement. Into the resulting expanded surface geometry an appropriate depression can now be designed to accommodate the bottle by defining two overlapping rectangular pockets. The secondary component surfaces for these two pockets are cylindrical; one into which the main body of the bottle can be seated and one in which the cap of the bottle can rest. These two feature constructs are shown in Fig. 8.8(b,c). This illustrative design modification required about 10 minutes to complete (90% of that time was spent rendering the surface iterates).

We remark here that the specular reflections in these and all other shaded renderings displayed in this paper are a consequence of the faceted triangular approximations used in the rendering process and not a result of any lack of smoothness in the underlying surface geometry.

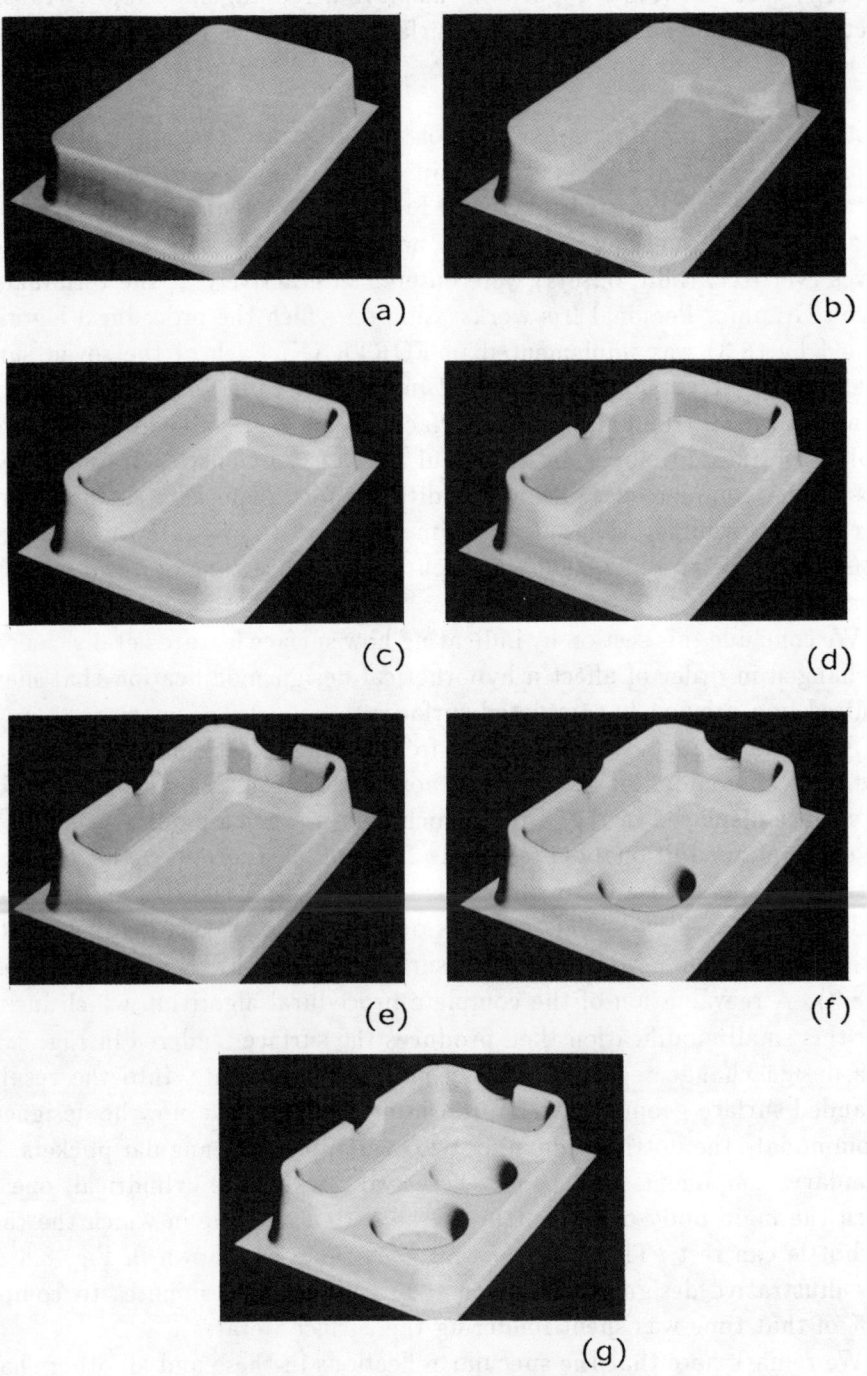

FIG. 8.7. *Seven-step design of contact lens cleaning kit.*

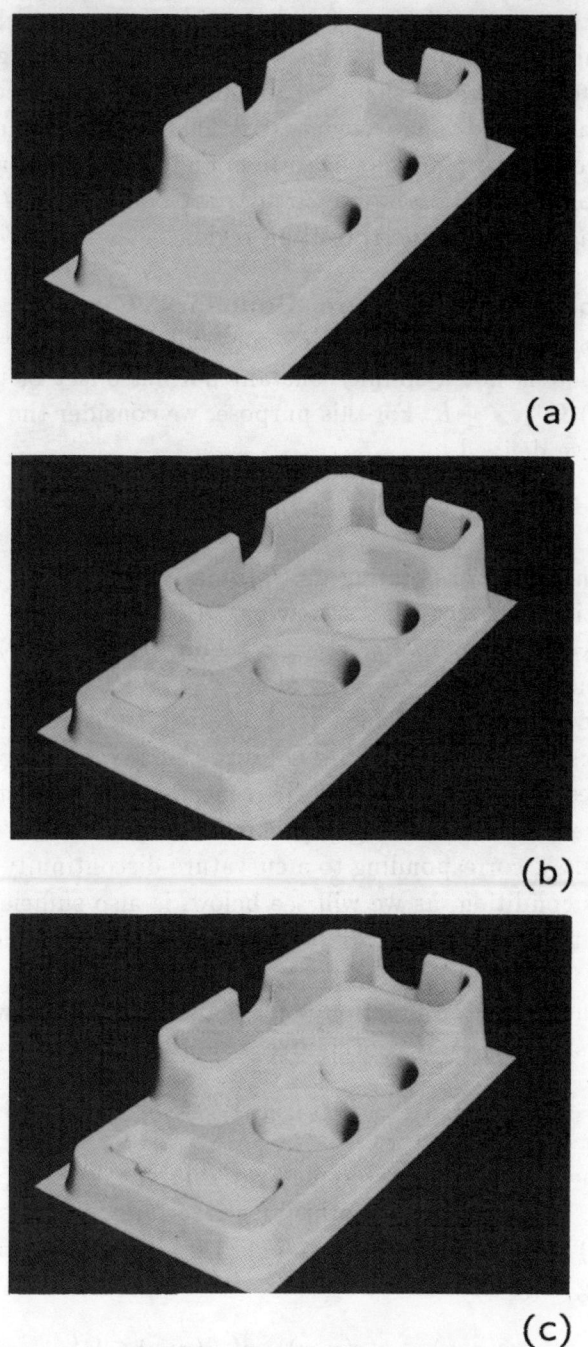

(a)

(b)

(c)

FIG. 8.8.   *Modification of contact lens cleaning kit design.*

8.3. Some Extensions and Enhancements

The main elements of the feature-based construction outlined in §8.2 are the transition function defined in §8.2.4 and the blending approach given by (8.2) and (8.3). While this approach defines an effective modeling tool for many functional surfaces, increased shaping capability can be realized in the method by making certain simple enhancements to it. These enhancements relate, first, to defining more general plan view feature boundary curves than those derived from constant offsets of straight line circular arc constructions and, second, to modifying surface shape in the transition region.

8.3.1. Variable Offset Feature Boundary Curves.
The design of certain features in a functional surface may call for an outer curve (denoted C_0 in §8.2.1) that is not a simple constant normal offset of the inner curve, $C_1 : (x(s), y(s)), 0 \le s \le L$. For this purpose, we consider the variable normal offset outer curve defined by

$$(8.4) \qquad C_0 : (x(s), y(s))^T + D(s)\hat{\eta}(s), \quad 0 \le s \le L$$

where $\hat{\eta}(s)$ is the outward pointing unit normal vector to C_1, and $D(s) > 0$ is a smooth function of arclength s satisfying $D(0) = D(L)$ and $D'(0) = D'(L)$. To avoid cusps and loops in C_0, we also require that $1/D(s) > \kappa(s)$, where $\kappa(s)$ is the signed curvature (see [3]) along C_1. We continue to assume that C_1 is composed of smoothly joined straight lines and circular arcs since this construction is sufficiently flexible to capture the shape of most feature inner boundary curves of practical interest. This means that in order for C_0, defined by (8.4), to have a continuous tangent, we must assume that $D'(s_0) = 0$ at each arclength value, s_0, corresponding to a curvature discontinuity along the inner curve C_1. This condition, as we will see below, is also sufficient to guarantee that $\nabla\Phi$ is continuous for the variable offset transition functions constructed as follows.

To define the transition function Φ for points $(x, y) \in \Omega^T$ when C_0 is given by (8.4), we first define local coordinates $\rho = \rho(x, y), s = s(x, y)$ satisfying

$$(8.5) \qquad (x, y)^T = (x(s), y(s))^T + \rho\hat{\eta}(s).$$

Next, we define the normalized radial variable $r(x, y) = \rho(x, y)/D(s(x, y))$ and, as before, let $h(r)$ be a smooth univariate function satisfying end point conditions $h(0) = 1, h'(0) = h(1) = h'(1) = 0$. The transition function Φ is given according to (8.1) with

$$(8.6) \qquad \Phi(x, y) = h(r(x, y)) \quad \text{if} \quad (x, y) \in \Omega^T.$$

The transition function, defined in Ω^T by (8.6), can be shown to satisfy:

$$(8.7) \qquad \nabla\Phi = h'(r(x, y))\left\{ \frac{\hat{\eta}(s)}{D(s)} - \frac{D'(s)\rho\hat{t}(s)}{D^2(s)(1 - \rho\kappa(s))} \right\}, \quad (x, y) \in \Omega^T,$$

where $\hat{t}(s)$ is the unit tangent vector to C_1 and $\kappa(s)$ is the signed curvature of C_1. From (8.7) it follows that $\nabla\Phi$ is continuous whenever $\kappa(s)$ is continuous. If $\kappa(s)$ has a finite jump discontinuity at a point s_0, then $\nabla\Phi$ will be continuous provided $D'(s_0) = 0$. Since we are assuming here that C_1 is comprised of circular arcs smoothly joined to straight lines, we must require that $D'(s) = 0$ at each parameter value s that locates a junction point. Geometrically, this amounts to requiring that, at each such joint, the outer curve has a tangent that is parallel to the inner curve's. If this proves too restrictive, then we must depart from the arc-line-arc construction and modify C_1 so that $\kappa(s)$ is continuous. Although we chose not to elaborate the details here, this can be done by smoothly "patching" a quintic Hermite polynomial curve across each junction point which then produces a geometric curve very close to the original curve C_1 but which now has continuous curvature for all values of arclength.

In §8.2.4 we suggested that $h(r)$ be a cubic Hermite polynomial. Other choices might be made to provide some shape control in Ω^T. These include higher-order polynomials, piecewise polynomials, or arcs of circles (or ellipses) connected smoothly by straight lines. An arc-straight-arc definition for $h(r)$ permits construction of a transition function Φ which, in Ω^T, consists of two fillets connected by a surface whose angle with the xy-plane is constant. This particular construction is important because it can be used for many surface features to define a transition surface that closely approximates a type of blending surface often called for in engineering practice: a *lateral filleted wall*. Specifically, if the component surfaces S_0 and S_1 defined in §8.2.1 are planes parallel to the plan view plane, then $z = g_1(x,y)$ defined by (8.2) is simply a scaled version of $z = \Phi(x,y)$. Although this is not generally the case, it is usually true that the component surfaces are gently curved, nearly parallel to the plan view, and approximately constant offsets of each other. Under these circumstances, the simple behavior of the transition function will be inherited, in an approximate sense, by the surface defined by (8.2).

If the surfaces S_0 and S_1 are rapidly changing in the transition region Ω^T, or if they intersect in Ω^T, it may happen that (8.2)'s shape inside Ω^T may not be well predicted by the shape of $z = \Phi(x,y)$. Then, it may be necessary to change the shape of (8.2) inside Ω^T. Shape modification can be achieved by the approach described below.

8.3.2. Generalizing the Blending Formula. The approach that we propose adjusting surface shape in the transition region involves the concept of a modifying function such as was proposed in [4]. Briefly, we modify (8.2) to assume the form

(8.8) $$g_1 = (1 - \Phi)f_0 + \Phi f_1 + \Psi_m$$

where the modifying function, Ψ_m, is smooth and nonzero only in the transition region Ω^T. The basic idea in this approach is to use the first two terms in (8.8) to achieve the smooth blend of surface S_0 with surface S_1 along the designated feature boundaries, and then adjust Ψ_m to impart particular shape characteristics to the surface within the transition region.

Although there is considerable flexibility in choosing Ψ_m, one that is particularly natural derives from making a more general interpretation of the blending formula (8.2). The interpretation arises by noting, as shown below, that the expression (8.2) may be viewed as a special case of a Bernstein polynomial in the variable Φ.

To generalize our interpretation of (8.2) we first recall (see [5]) that a Bernstein polynomial of degree n is constructed by forming a linear combination of the Bernstein basis functions

$$(8.9) \qquad B_{i,n}(u) = \binom{n}{i} u^i (1-u)^{n-i}, \quad i = 0, 1, \cdots, n.$$

In the special case where $n = 1, B_{i,1}, i = 0, 1$ are linear polynomials in u. This means that (8.2) can be expressed as

$$(8.10) \qquad g_1(x,y) = \sum_{i=0}^{1} v_i B_{i,1}(\Phi(x,y))$$

where $v_0 = f_0(x,y)$ and $v_1 = f_1(x,y)$. We generalize this by simply increasing the degree, and hence the number of associated Bernstein basis functions. In particular, we form the expression

$$(8.11) \qquad g_1(x,y) = \sum_{i=0}^{n} v_i B_{i,n}(\Phi(x,y))$$

where $v_0 = f_0(x,y), v_n = f_1(x,y)$, and $v_1, \cdots, v_{n-1}$ are coefficients, perhaps depending on (x,y), to be determined. We note first that (8.11) provides both a blending capability similar to (8.2) and, when $v_i, i = 1, \cdots, n-1$ are not all zero, a modification capability along the lines sought in (8.8). This is seen by reordering the terms of (8.11) as follows:

$$(8.12) \qquad \sum_{i=0}^{n} v_i B_{i,n}(\Phi) = v_0 B_{0,n}(\Phi) + v_n B_{n,n}(\Phi) + \sum_{i=1}^{n-1} v_i B_{i,n}(\Phi).$$

The first two terms in (8.12) provide the required blending capability. The ability to modify this surface, solely within the transition zone, is provided by the remaining sum in (8.12). The function defined by this sum,

$$(8.13) \qquad \Psi_m = \sum_{i=1}^{n-1} v_i B_{i,n}(\Phi) = \sum_{i=1}^{n-1} v_i \binom{n}{i} \Phi^i (1-\Phi)^{n-i},$$

satisfies the requirements of the modifying function defined in (8.8). Smoothness of Ψ_m holds provided $\partial v_i / \partial x, \partial v_i / \partial y$ are continuous. The fact that (8.13) vanishes outside the transition region follows from the properties of $\Phi(x,y)$.

By adjusting the coefficients v_i in (8.13) we may modify the surface (8.12) inside the transition region only, without changing smoothness properties. We remark here that it is not necessary to use the modifying function Ψ_m given in (8.13) only with the associated (high degree) blending capability implied by (8.12). We may instead resort to (8.8) and retain the original blending approach. We advocate this approach because with $v_1 = v_2 = \cdots = v_{n-1} = 0$, (8.8) written here as

$$(8.14) \qquad g_1 = (1 - \Phi)f_0 + \Phi f_1 + \sum_{i=1}^{n-1} v_i \binom{n}{i} \Phi^i (1 - \Phi)^{n-i}$$

reduces to the original blending formula (8.2). This simple blending approach has proven to be very useful and appropriate in many circumstances and it is desirable to preserve the capability in a simple way.

One possible drawback of the modifying function appearing in (8.14) is the fact that each term in the sum has support over the entire transition region. This means that local shape adjustment can be accomplished only by changing all the coefficients v_i. To address this issue in a simple way we may proceed by formally replacing the Bernstein basis functions in (8.14) with a B-spline basis, [2], defined over the interval $[0, 1]$, with a given knot set ξ. For example, with kth-order splines in $S_{k,\xi}$, the blending formula (8.14) becomes

$$(8.15) \qquad g_1 = (1 - \Phi)f_0 + \Phi f_1 + \sum_i v_i N_{i,k,\xi}(\Phi).$$

Here, the sum is over those B-splines, $N_{i,k,\xi}$, which vanish at the end points $\Phi = 0, 1$.

8.4. Using Parametric Component Surfaces

In practice some the component surfaces will be defined parametrically (as Bézier or B-spline surfaces, for example) and instead of an explicit formula giving z as a function of x and y, the available description is of the form

$$(8.16) \qquad x = X(u, v), \quad y = Y(u, v), \quad z = Z(u, v),$$

where u and v are parameters ranging over a rectangular domain, $u \in [a, b], v \in [c, d]$. In (8.16) we assume that the coordinates (x, y, z) are relative to a given plan view orientation.

To use a component surface represented parametrically in the manner of (8.16) with our feature-based approach, it is necessary to duplicate the effect of an equivalent, explicit representation by numerical means. That is, given a point (x_0, y_0) in the plan view plane, we must find the z-value, z_0, so that (x_0, y_0, z_0) is on the given parametric surface. This means that we must first find corresponding parameter values $(u, v) \in [a, b] \times [c, d]$ which satisfy the equations

$$(8.17) \qquad \begin{aligned} x_0 &= X(u, v), \\ y_0 &= Y(u, v). \end{aligned}$$

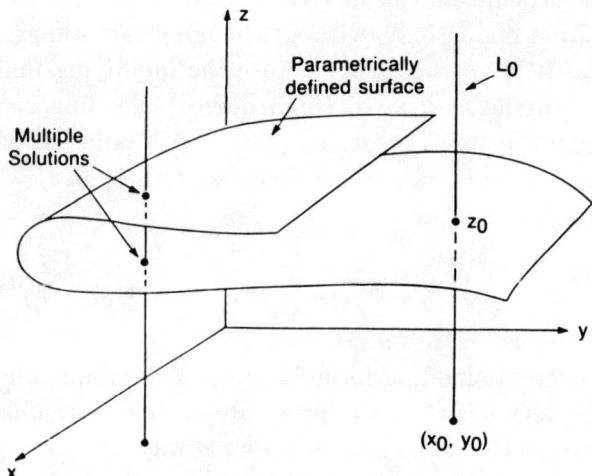

FIG. 8.9. *Parametric surface.*

Once this is done, the appropriate z-value, z_0, can be determined from the last equation in (8.16). The basic requirement is then to solve the system of equations (8.17). We have used a Newton iteration to solve the nonlinear system (8.17) for the example discussed in §8.5.

Before attempting to solve (8.17) we should gain assurance that a solution exists and that it is unique. The existence requirement is simply a demand that the parametric component surface extend sufficiently, or be properly positioned, to cover the appropriate regions in the plan view. The concern for uniqueness acknowledges the fact that parametrically defined surfaces may turn over on themselves, making multiple solutions of (8.17) possible. This is illustrated in Fig. 8.9. To deal with the uniqueness concern, the nonlinear problem (8.17) has to be reposed to include additional restrictions on admissible parameter values. In particular, we would have to restrict the solution search to a subregion, U, of the full parameter space — one that corresponds to the desired "layer" of the parametric component surface in question.

8.5. Examples

In this section we present four examples to show typical design sequences for realistic functional surfaces. All of the surfaces are designed using methods described in §8.2. The generalizations discussed in §8.3 were not required.

The order of the steps shown in each example can, in many cases, be varied. If the features being designed do not overlap, then the process is independent of order. If pockets do overlap, then the last pocket constructed will dominate in the overlap region. This point is evident in Example 1. Other aspects of the capabilities of the feature-based approach are shown in the remaining examples.

Example 2 illustrates a novel means of defining needed secondary surfaces as iterates in the recurrence (8.3). The use of parametric surfaces is highlighted in Example 3. Finally, the utilization of various input parameters as a natural means of providing shape control is noted in Example 4.

Example 1.   We first illustrate a 9-feature automobile inner panel design effort. The first six steps and the ninth step are shown in Fig. 8.10 (a–g). For this exercise, all component surfaces are parallel or inclined planes. The total design effort required inputting 201 numbers: the x and y coordinates of 54 vertices, 54 plan view radii values, 9 constant offset values to specify the boundary curves of the 9 pockets which comprise the features of this functional surface, and 30 numbers which provide math descriptions of the 10 planar component surfaces $S_0, S_1, \cdots, S_9$. The interactive entry of this data requires less than 15 minutes at the keyboard. Evaluation of the surface on a 200×200 rectangular grid for shaded rendering purposes requires less than a minute of terminal time at a Silicon Graphics Personal Iris workstation.

Figure 8.10(h) displays the actual sheet metal panel from which we extracted estimates of the design input information needed to execute the recurrence in (8.3). It is estimated that using a parametric patch-based approach to represent the surface in Fig. 8.10(h) would require over 200 separate surface patches and 3 man-days of effort.

Example 2.   It is possible to use our feature-based approach to define more complicated secondary surfaces than can conveniently be represented by analytic explicit or parametric expressions. In particular, by properly planning the blending steps, we can arrange to have such secondary surfaces emerge as intermediate surfaces, $z = g_i(x, y)$, for later use in a given feature-based design effort.

A simple illustration of this occurs in the design of the automobile engine valve cover shown in Fig. 8.11(f). To design and render this surface using (8.3) we used a six-step recurrence. Five of these steps are shown in Fig. 8.11(a–e). The first three steps are used to define the secondary surface that will represent the top of the valve cover.

In step 1, a cylindrical surface $(z = f_1(x, y))$ is blended to the original planar base surface $(z = f_0(x, y) = 0)$. In steps 2 and 3 conical surfaces $(z = f_2(x, y)$ and $z = f_3(x, y))$ are blended to each end of the cylindrical surface. The result is the intermediate surface $z = g_3(x, y)$ shown in Fig. 8.11(c).

In step 4 (not shown) the base surface is reintroduced by blending the plane $z = 0$ to the current primary surface. For this step, the inner curve is chosen so that its interior contains all features designed so far. In step 5, shown in Fig. 8.11(d), the output of step 3 $(z = g_3(x, y))$ is introduced as a secondary surface and blended to the current primary surface. In more complicated designs, there may be additional design steps separating the reinitialization

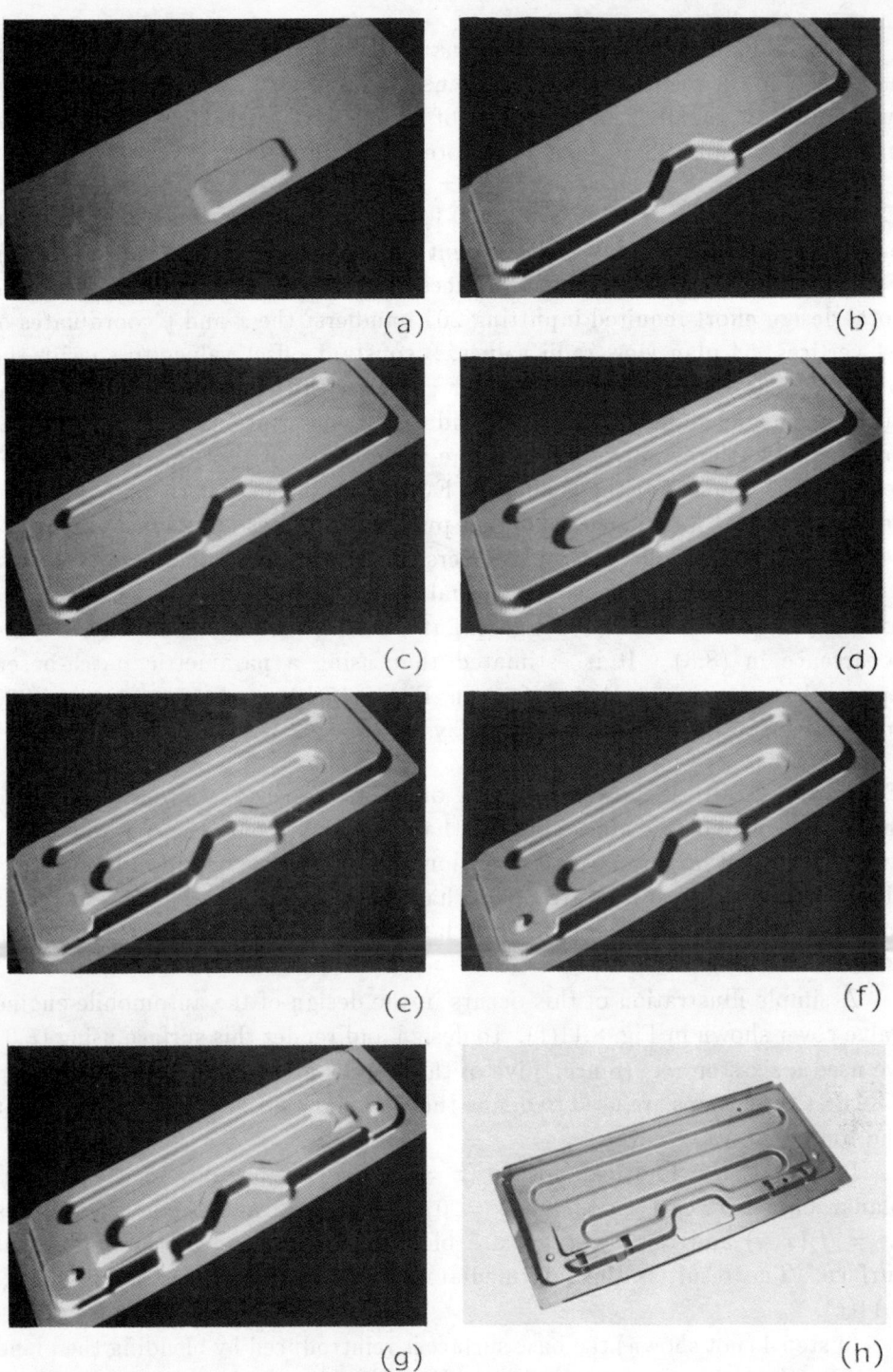

FIG. 8.10. *Nine-pocket design of an automobile inner panel. Seven of the nine design steps are shown.*

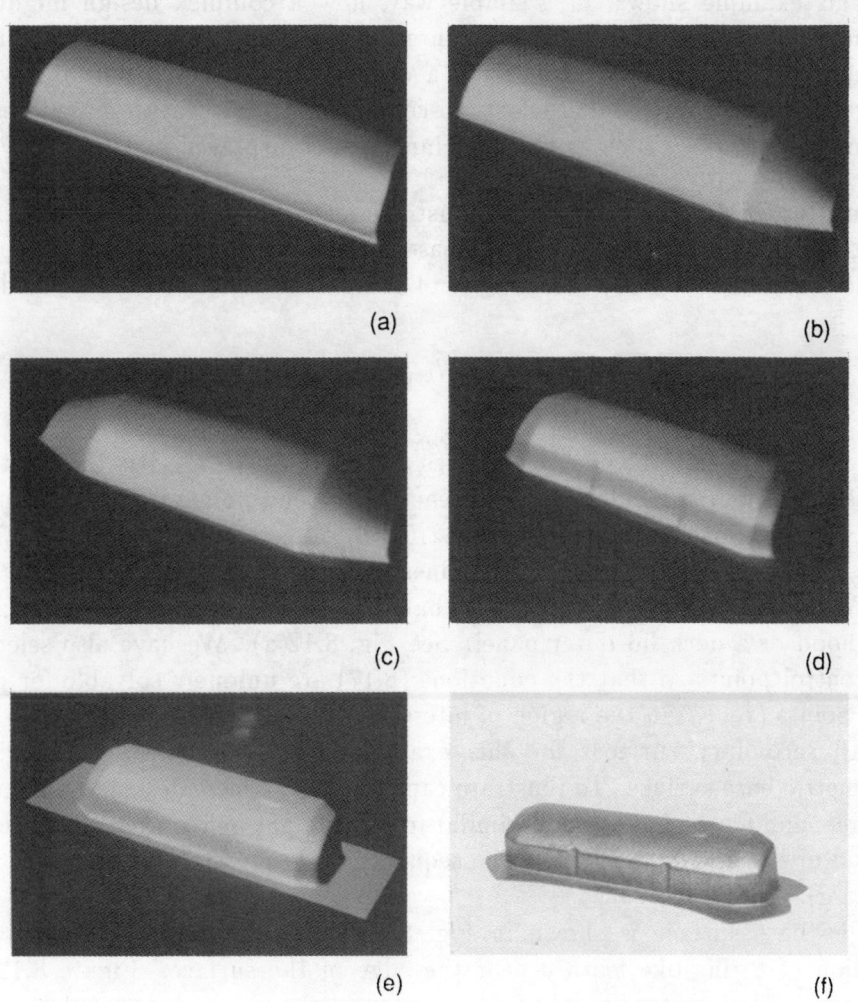

(a)

(b)

(c)

(d)

(e)

(f)

FIG. 8.11. *Recursive design of an engine valve cover.*

(step 4) and the utilization of the previously generated intermediate surface (step 5). In cases such as this, where these steps are adjacent, the steps may be combined by means of a surface reversal technique. At step 4 we would merely interchange the roles of the current primary surface ($z = g_3(x, y)$) with the secondary surface ($z = f_4(x, y) = 0$) and input plan view information that would normally be given in step 5. The last step adds the small, flat-topped pocket as shown in Fig. 8.11(e).

This example shows, in a simple way, how a complex design might be decomposed into several smaller feature-based designs, and then combined to represent the final surface. Such a decomposition would permit several designers to conduct parallel efforts to contribute to the feature-based design of a complex surface such as the door inner panel displayed in Fig. 8.3.

Example 3. In this example we illustrate feature-based design capabilities using parametric surfaces. Here, the base surface is a tensor product B-spline surface of the form (8.16). The vector of component functionals is defined by

$$P(u, v) = \sum_{j=1}^{9} \sum_{k=1}^{9} P_{jk} N_{j,4,\xi}(u) N_{k,4,\xi}(v), \quad (u, v) \in [0, 1] \times [0, 1],$$

where $P_{jk} \in R^3$ are the control vertices and $\xi = \{\xi_1, \cdots, \xi_{13}\}$ is the knot set consisting of four coincident knots at each of the end points 0 and 1, together with five interior knots located at $0.2, 0.3, 0.5, 0.7$, and 0.8. The functions $N_{j,4,\xi}$, $j = 1, \cdots, 9$ are the cubic B-splines spanning $S_{4,\xi}$.

We have chosen control points so that the surface assumes a shape similar to a hood or a deck lid outer panel. See Fig. 8.12(a). We have also selected the control points so that the equations (8.17) are uniquely solvable for plan view points (x_0, y_0) in the region of interest.

All secondary surfaces for this example are z-translates of the given parametric base surface. To illustrate capabilities, we have designed a number of three- and four-sided pockets, similar to what might be required to represent a hood or deck lid inner panel. The sequence of design steps is shown in Fig. 8.12(a-g).

The base surface is shown in Fig. 8.12(a). Figure 8.12(b,c) show the addition of a ring-like feature near the edge of the surface. Figure 8.12(d–g) show the design steps to add first a trapezoid-shaped pocket and then three triangular-shaped pockets. If this were an actual hood inner panel, the interiors of each of these pockets would be cut out, leaving a network of stiffening ribs that would mate with the underside of the corresponding outer panel.

It is important to note that, aside from the increased computational effort needed to evaluate parametric surfaces at plan view locations, there is no essential difference in the manner of implementation.

Example 4. Our last example illustrates the design the functional surface geometry of the plastic frozen food serving plate displayed in Fig. 8.3(c). This

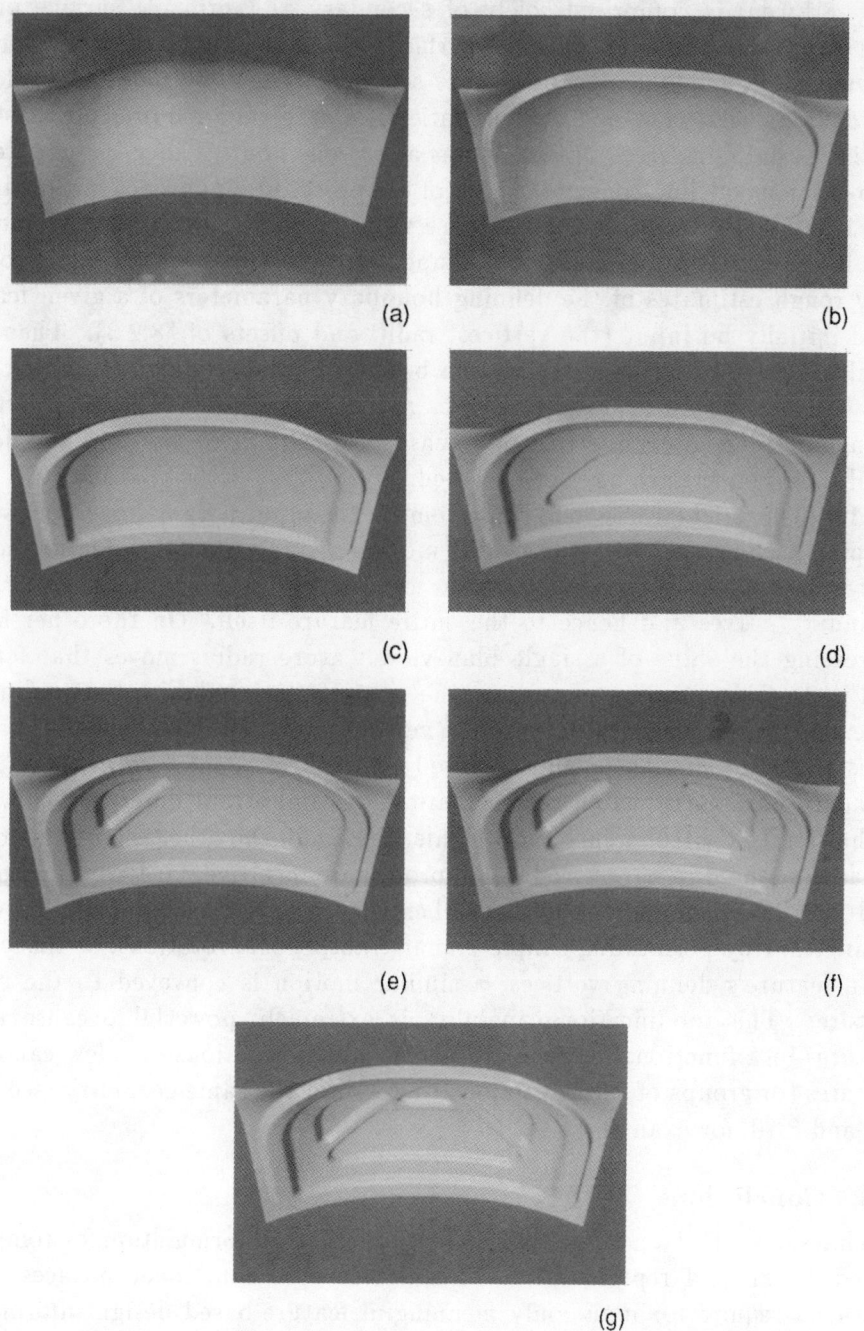

(a)

(b)

(c)

(d)

(e)

(f)

(g)

FIG. 8.12.   *Example of feature-based design using parametric component surfaces.*

particular exercise involved the construction of an 8-step recurrence for (8.3). The results of four of these steps are shown in Fig. 8.13(a–d). The final surface (Fig. 8.13(d)) is comprised solely of secondary and primary surface planes connected by blended transitional surfaces that are univariate cubic Hermite polynomials along radial pocket cross sections. The time required to design the defining plan view pocket boundaries (i.e., the time required to construct the input data discussed in §8.2.3) was about one hour. This necessary design step is the most time-consuming part of the procedural approach outlined here and it is greatly facilitated by the ease with which feature surface geometries can be interactively modified and shaped with this method. That is to say, only rough estimates of the defining boundary parameters of a given feature need initially be input (the vertices, radii, and offsets of §8.2.3). Then, the resulting pocket surface geometry can be rapidly rendered and evaluated, and simple modification operations such as movement of selected feature vertices, change in selected radii, etc., can be easily carried out to fine tune the design until a desired surface has been created.

In this particular design exercise some of the input parameters were used as shaping parameters. For example, by moving a single plan view feature corner vertex in a given direction, a similar motion is imparted to the associated boundary curves and hence to the entire feature itself. On the other hand, increasing the value of a single plan view feature radius moves that feature and its boundary curves away from the vertex associated with that radius. In a certain sense, then, the plan view vertices and radii play surface shaping roles that are similar to those played by control points and weights in a standard parametric patch representation. An important difference, however, is that unlike control points and weights, these and the other shape controlling parameters used in our procedural approach are intuitive, far fewer in number, and they have meaningful physical and engineering significance. It is also worth noting that by performing simple planar translations, rotations, or reflections on a feature's defining vertices, a similar motion is conveyed to the entire feature. This modification capability is extremely powerful because many features in a functional surface are simple transformations of a few canonical features (or groups of features) appearing within the same geometry (see Figs. 8.3 and 8.10, for example).

8.6. Conclusions

We have described a new feature-based mathematical formulation for computer aided design and representation of multi-featured functional surfaces. The methods require for input only meaningful feature-based design information which an engineer or designer naturally uses to describe the geometry of a particular surface. Tests involving the design of realistic functional shapes have shown the methods to be effective, efficient, and well suited to the design requirements of such shapes.

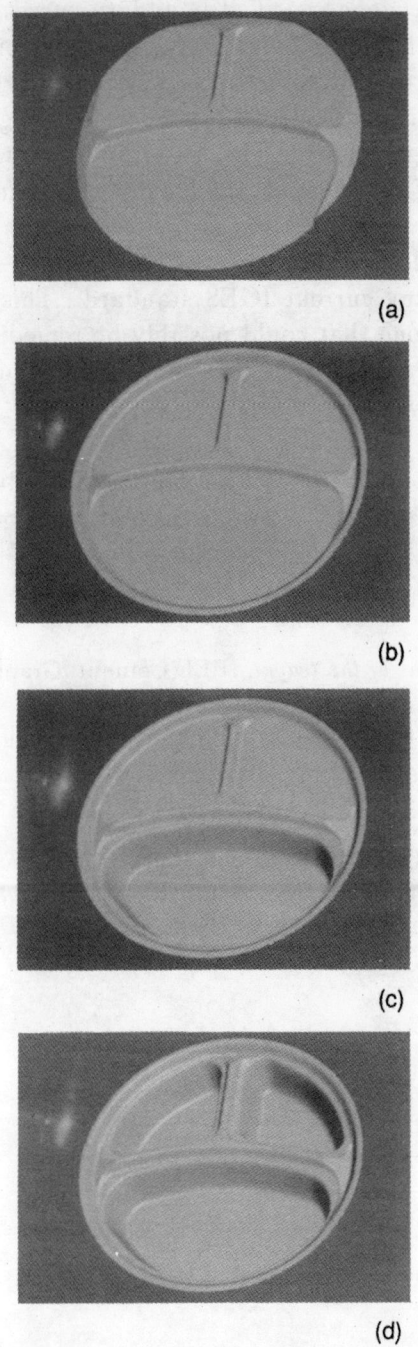

(a)

(b)

(c)

(d)

FIG. 8.13. *Recursive design of frozen food dish. Four of eight steps are shown.*

The new approach is motivated by viewing the process of functional surface design as one of "surface assembly," in which a number of known component surfaces are combined, along specified boundary curves, with a given base or primary surface. With the aid of a novel strategy for recursively blending the component surfaces to the base surface, we are able to quickly and accurately build complicated multi-featured surface designs through a sequence of relatively simple operations. The smoothness of the output surfaces is rigorously achieved. This is important for down stream processing functions such as NC machining.

The surfaces we design through (8.3) are defined procedurally and are not compatible with any current IGES standard. This, we believe is only a temporary drawback, one that could possibly be remedied, in the short term, by fitting a standard surface to the procedurally defined surface.

Acknowledgments

We would like to thank Giles Ross for his support and advice in the formulation and implementation of this work. We are also grateful to Dr. Paul Besl for his help in generating the shaded images appearing in this paper.

References

[1] F. Crow, *The origins of the teapot*, IEEE Comput. Graphics Appl., 7 (1987), pp. 8–19.

[2] C. deBoor, *A Practical Guide to Splines*, Prentice-Hall, Englewood Cliffs, NJ, 1978.

[3] M. P. doCarmo, *Differential Geometry of Curves and Surfaces*, Prentice-Hall, Englewood Cliffs, NJ, 1976.

[4] J. P. Duncan and G. W. Vickers, *Simplified method for interactive adjustment of surfaces*, Comput. Aided Des., 12 (1980), pp. 305–308.

[5] G. Farin, *Curves and Surfaces for Computer Aided Design*, Academic Press, Boston, MA, 1988.

Topological Considerations in the Interpolation of Contour Curves

Alan K. Jones

9.1. Introduction

The contour interpolation problem may be described as follows. Reconstruct a surface embedded in space, given only a set of discrete curves, which may be assumed to lie in parallel planar cuts through that surface. Typical applications arise in medical and geological imaging and in computer vision. In all of these cases, the surface may be taken to be the boundary of a region in R^3, say a tumor or an orebody, and hence a closed surface. The desired result is a representation of the boundary surface as a union of planar polygons, preferably triangles, suitable for high-speed graphical rendering.

Difficult geometrical problems arise in contour interpolation, principally those of contour matching. That is, which points on two adjacent contours are to be blended? The problem is exacerbated in branching situations where each contour may have multiple components, for then it is necessary to know which components are connected by surface and which are not. In some situations, for example, computer vision applications where one object partially occludes another, or cartographic applications where the input contour system is incomplete, it may be necessary to infer the existence of additional contour components in order to reconstruct the correct surface. Details can be found in the large existing literature in medical imaging and computer vision, of which [1], [2], [4]–[6], [11]–[13], [21], [23] are merely a sample.

This paper takes a slightly different view of the problem, in which the desired end result is not a shaded graphics rendering of the surface, but rather a model from which intermediate contours can be extracted, varying smoothly between those that were input. As discussed in §9.4 below, these techniques are expected to be useful in, and were partly motivated by, problems in finite difference grid generation for computational fluid dynamics.

To make contour extraction as fast as possible, the surface is modeled not as a collection of flat triangles, but as a collection of smooth, parametrically rectangular patches, each joined smoothly to its neighbors. Each contour curve is then representable as a union of isoparameter curves, one for each patch that

169

it crosses. To distinguish the new, intermediate contours from the original inputs, they will henceforth be referred to as *pseudocontours*.

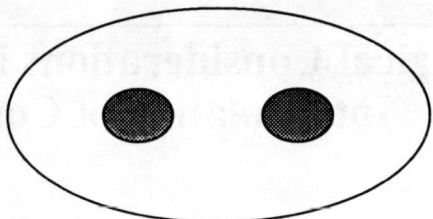

FIG. 9.1. *A multiply connected domain.*

In the presence of branching, singular points will arise in the grid of rectangular patches, where the number of patches meeting at a corner is other than four. A simple case is illustrated in Fig. 9.1. The area to be filled with rectangular parametric patches is the unshaded region, a disk with two holes. As shown in §9.2, this is the typical configuration near a branch or saddle point. The reader may attempt as an exercise to fill this region with nondegenerate rectangular patches, meeting four at a corner. A proof that this is impossible will be presented in §9.3. An alternative, in which eight patches meet at a common vertex, is shown in Fig. 9.2.

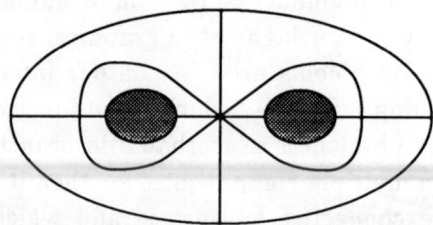

FIG. 9.2. *A block design.*

As is well known, such points require special techniques to preserve geometric smoothness of the surface. There is an extensive literature on this topic as well. See, for example, [8], [17], or [20]. Since parametric smoothness is not possible, pseudocontours as defined above will display derivative discontinuities near these singular points. The discontinuities may be smoothed out by replacing isoparameter curves by true planar intersection curves in these exceptional regions.

However, the focus of the present work is not on the details of construction of these interpolants, but rather on the underlying topological question of how the rectangular patches should be arranged and connected so as to produce topologically correct pseudocontours. Three-dimensional analogues of these techniques exist, as reported in [10], and the exposition in the present work is intended in part to provide simple motivating examples for these later complexities.

The remainder of this paper is organized as follow. Section 9.2 provides a general framework for classifying the singularities that arise in contour maps. Section 9.3 provides a parallel classification of singularities in arrays of parametrically rectangular surface patches. Section 9.4 illustrates the relationship between the two, exploring the impact of choices in patch network topology on the topology of pseudocontours. Section 9.5 discusses applications to the classic problem of closed surface reconstruction.

In some problems of this sort, the underlying surface is the graph of a function over a bounded domain, as in traditional two-dimensional cartographic applications, and is thus a surface with boundary. See, for example, [16]. This introduces additional considerations, which are discussed in §9.6. Further applications, including grid generation in domains with nonsmooth boundaries, as described in [9], and, of course, the three-dimensional cases, will be reported in future work.

9.2. A Review of Morse Theory

The classic mathematical tool for building up the shape of a closed surface from data along parallel sections is given by Morse theory, introduced in [15]. This chapter will provide a brief introduction to that material. Although the conclusions that will be drawn in the two-dimensional case may not seem to merit the investment in such heavy machinery, it provides a useful framework within which to generalize these results to later constructions in higher dimensions. A good elementary reference is Part I of [14]. See also the first essay in [18].

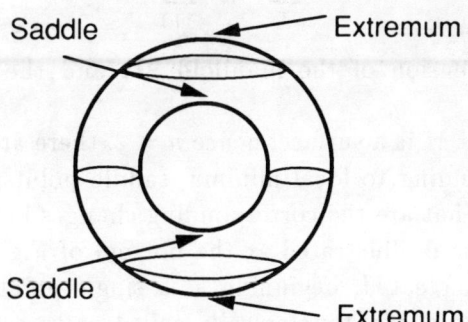

FIG. 9.3. *Morse function on a torus.*

For the following discussion, Fig. 9.3 is the canonical example. It illustrates a manifold M, here just a torus, embedded in space, with a real-valued function f defined on M, which in this case is just the projection on the z coordinate axis, or, equivalently, the height above the xy plane. Each of the given parallel plane sections of M is $f^{-1}(c)$, for some c.

Assume that M is a smooth surface, so f is at least C^1. The *critical points* of f are defined to be the points x of M at which the gradient of f vanishes. In this example, they are the four points on the torus with horizontal tangent planes. The function f is said to be a Morse function if at each critical point x of f, the Hessian of f, which is a 2×2 matrix denoted $h(x)$, is nondegenerate. Any function f on a smooth manifold can be approximated arbitrarily closely by a Morse function. In the language of Thom, we can say that the Morse functions are structurally stable. So, in practice, we can assume that all height functions encountered in practice are Morse, within experimental error.

The fundamental result of Morse theory deals with the sublevel sets of the form $M_c = f^{-1}(-\infty, c]$, i.e., the set of points in M at or below level c. It asserts firstly that if $b < d$, then M_b is topologically equivalent to M_d provided there is no critical point x of f such that $b \leq f(x) \leq d$. The proof proceeds by showing that the new surface added between levels b and d is a cylindrical collar with base $f^{-1}(b)$ that can be deformed smoothly into M_b without changing the topology of the latter. In particular, each intermediate contour $f^{-1}(c)$ is a slice through this collar, topologically equivalent to $f^{-1}(b)$. For the purposes of this work, this is an important observation. It establishes that contours do not change topological type except at critical values of f, i.e., values of the form $f(x)$ for some critical point x.

The deeper half of Morse theory explains how M_b and M_d are related when the new part of the surface does contain exactly one critical point x_0, where $f(x_0) = c_0$. (Nondegenerate critical points are necessarily isolated, so each can be treated separately.) The possibilities are completely classified by the index $\lambda(x_0)$, which is the number of negative eigenvalues of the matrix $h(x)$. Then, up to second order, we can approximate f by

$$c_0 - \sum_1^\lambda y_i^2 + \sum_{\lambda+1}^n y_i^2,$$

where n is the dimension of the manifold M, and the y_i are appropriate coordinates on M.

In the case where M is a surface, hence $n = 2$, there are three possibilities, $\lambda = 0, 1, 2$, corresponding to local minima, saddle points, and local maxima of f, respectively. What are the corresponding changes in topological type for the contours? In case 0, illustrated at the bottom of Fig. 9.3, a new contour component has been created, beginning as a single point when $c = c_0$, and expanding to a closed curve, topologically eqivalent to a circle, when $c > c_0$. Case 2, illustrated at the top of the figure, is the reverse process, where a circular contour component at $c < c_0$ shrinks to a point at $c = c_0$, and then disappears. Finally, case 1 is a branch point, where one contour component has split into two, as in the lower example, or two have fused into one, as in the upper example.

Thus, the topology of the contours of a surface is completely determined by the positions and indices of the critical points of the height function. Unfortunately, the converse is not true. Given a typical, probably incomplete set of contours, there is no purely topological way even to compute the indices of the intervening critical points, much less their locations. The topology

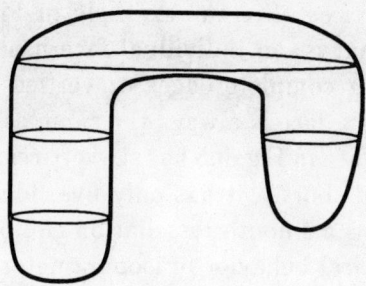

FIG. 9.4. *An alternative surface reconstruction.*

of a contour curve on a closed surface is that of a disjoint union of circles; all we can do is to count the circles. When a contour at level d has one more component than that at b, this could mean that a saddle lies between them, or perhaps a relative minimum, and similarly for the opposite case. See, for example, Fig. 9.4, which has the same contour set as Fig. 9.3, but is an entirely different surface. This emphasizes that the question of which contour components to match at different levels must be resolved by some different means. The references cited in §9.1 describe a variety of different algorithms for making these decisions.

9.3. Singularities of Block Decompositions

We have now related the topology of contours of a smooth map f to the singularities of f. The next goal is to characterize the topology of the intermediate pseudocontours generated as unions of isoparameter curves in a piecewise parametric surface. To that end, we investigate the possible forms of irregularity in the structure of an assemblage of parametric patches. The following analysis is a slight generalization of results described in [9], and is related to the earlier unpublished work [3]. It is included here to make the exposition more self-contained. Special cases, at least those relating to the Euler characteristic of closed surfaces, seem to be well known in the folklore of grid generation. See, for example, [19].

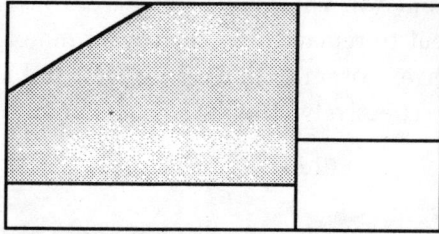

FIG. 9.5. *A block design.*

A *smooth tessellation* of a domain X is a subdivision of that domain into disjoint *faces*, which are the smooth images of open polygons, joined along disjoint *edges*, which are the smooth images of open line segments, joined in

turn at points called *vertices*. See the example in Fig. 9.5. Each face must
be simply connected. That is, an individual face has no holes. Furthermore,
adjacent faces meet along complete edges or vertices. Thus, the number of
edges on the boundary of a face is always the same as the number of vertices.
For example, the shaded face in Fig. 9.5 has six vertices and six edges, although,
in a sense to be described shortly, it has only five sides.

A *block design* for X is a smooth tessellation equipped with two functions,
α and β, to specify the local behavior of isoparameter curves near vertices.

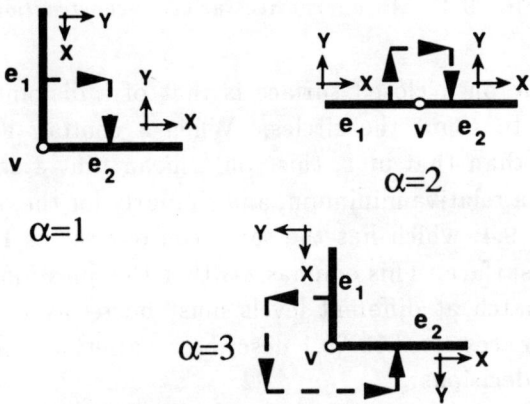

FIG. 9.6. *Computing angle counts.*

At each vertex v on a face f, define $\alpha(v, f)$ to be the number of grid
angles at v in f, defined as follows. Let e_1 and e_2 be the edges of f meeting
at the vertex v. Near each edge, define a local (x, y) coordinate system
composed of isoparameter curves, with x parametrically parallel to the edge
and y parametrically normal to it, pointing into the face. Then $\alpha(v, f)$ is the
number of switches of isoparameter curve species required to move from the
$-y$ (outward normal) direction along e_1 to the $+y$ (inward normal) direction
along e_2. Each such switch is a *grid angle*. Figure 9.6 illustrates this procedure,
showing that a typical convex corner has $\alpha = 1$, a concave corner $\alpha = 3$, and
a vertex which is not a corner at all has $\alpha = 2$. Other values of α can arise,
for example at cusps, but will not concern us here.

It is often convenient to replace α by the corner index $\kappa(v, f) \equiv 2 - \alpha(v, f)$,
and to characterize convex corners, concave corners, and noncorners as vertices
where $\kappa = +1, -1, 0$, respectively. Define

$$\sigma(f) = \sum_{v \in f} \kappa(v, f)$$

to be the *number of sides* of the face f. The boundary of f can be decomposed,
not necessarily uniquely, into $\sigma(f)$ sides, each of which is a string of consecutive
edges $e_0, \cdots e_{k+1}$ meeting at vertices $v_1, \cdots, v_k$, such that $\sum_{j=1}^{k} \kappa(v_j, f) = 0$.
See Fig. 9.7 for examples.[1]

---

[1]The author is indebted to F.B. Holt for useful conversations on this topic.

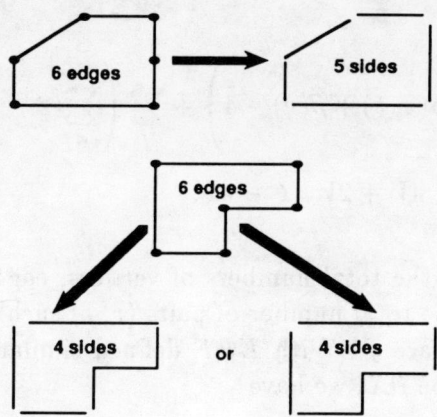

FIG. 9.7. *Finding sides of faces.*

The function β is defined by analogy with α to specify behavior near the boundary of X, which we denote ∂X. If v is an interior vertex of X, $\beta(v) = 0$. If $v \in \partial X$, $\beta(v)$ is the number of grid angles at the vertex v external to the domain X. Thus if the boundary is parametrically smooth at v, $\beta(v) = 2$; if the boundary looks like a convex corner at v, typically $\beta = 3$; at a concave corner, $\beta = 1$. Define the *boundary score* to be, intuitively, the net number of convex corners inside the domain on the boundary. Formally, it is defined as

$$\Omega_B = \sum_{v \in \partial X} [\beta(v) - 2].$$

For applications in the current work, $\beta \equiv 2$ along ∂X, and $\Omega_B \equiv 0$.

Define the *singularity count* functions as follows.

If f is a face, set $\delta(f) = \sigma(f) - 4$, the net number of excess sides in the face. A value of $\delta \neq 0$ means a nonrectangular patch.

If v is a vertex, set $\gamma(v) = \left(\sum_{v \in f} \alpha(v, f) \right) + \beta(v) - 4$, the net number of excess grid angles assigned at v. If $\gamma = 0$, then the local (x, y) coordinate systems in the patches meeting at v (including, because of the β term, those beyond the boundary of X) can be hooked together to give a global (x, y) system in a neighborhood of v. If $\gamma \neq 0$, this cannot be done.

The *score* of the block decomposition is defined to be

$$\Omega = \sum_{v} \gamma(v) + \sum_{f} \delta(f),$$

which is the total net number of singularities in the grid structure defined by the local (x, y) coordinate systems.

THEOREM 9.3.1. $\Omega = \Omega_B - 4\chi(X)$, *where* χ *denotes the Euler characteristic.*

In particular, Ω is independent of the details of the blocking. It is a joint invariant of the topological type of X and the assignment Ω_B of parameter space angles to the boundary of X.

Proof.

$$\Omega = \sum_v \left(\sum_{v \in f} \alpha(v,f) + \beta(v) - 4 \right) + \sum_f \left(\sum_{v \in f}(2 - \alpha(v,f)) - 4 \right)$$

$$= \sum_v \beta(v) - 4V + 2V\#F - 4F,$$

where V, E, F denote the total numbers of vertices, edges, and faces, respectively, and $V\#F$ is the total number of pairs (v, f) such that v is a vertex on the boundary of the face f. With $E\#F$ defined similarly, and E_B denoting the number of edges on ∂X, we have

$$V\#F = E\#F = 2E - E_B,$$

because each edge separates precisely two faces, unless it is a boundary edge, in which case it meets only one face. Substituting into the first equation,

$$\Omega = \sum \beta(v) - 4V + 4E - 2E_B - 4F$$

$$= \sum_{v \in \partial X} (\beta(v) - 2) - 4(V - E + F)$$

$$= \Omega_B - 4\chi(X).$$

The only cases that will be needed in the present work are smooth blockings (i.e., $\Omega_B = 0$) of domains X = a disk with H holes. In this case, $\chi(X) = 1 - H$, so $\Omega(X) = 4(H - 1)$.

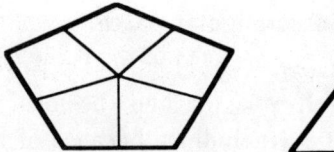

FIG. 9.8. *Face versus vertex singularities.*

As this theorem suggests, there is a duality between polygonal patches where $\delta \neq 0$ and vertices where $\gamma \neq 0$. Figure 9.8 shows explicitly how singularities of one type may be exchanged for singularities of the other. To emphasize this interchangeability, vertices with $\gamma = 3, 4, 5$, etc., are referred to as *triangular, rectangular, pentagonal*, and so on.

9.4. Blocking at Singularities

We begin with a discussion of the saddle point singularity. A generic example, illustrated in Fig. 9.1, is a region topologically equivalent to a disk with two holes. As shown in §9.3, the score $\Omega = +4$. We wish to fill the region with smooth images of rectangles, each of which has therefore at most four sides.

There is no advantage at this point to introducing degenerate rectangles, so we can assume $\sum \gamma(v) = 4$. The question then becomes, how best to allocate the nonzero γ values that sum to 4.

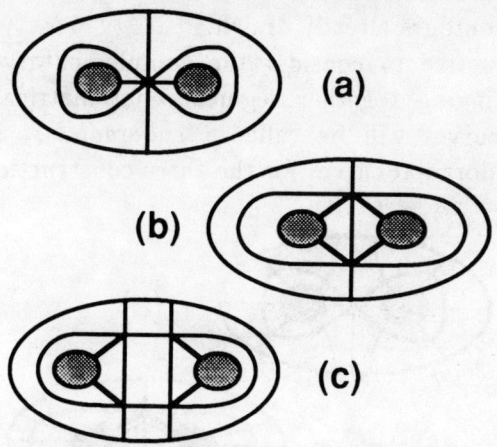

FIG. 9.9.   *Block designs at a saddle.*

Figure 9.9 shows a sequence of examples. In (a), the entire γ is concentrated at a single octagonal vertex. This is just a simplified version of Fig. 9.2. In (b), the octagonal vertex has been split into two hexagonal vertices, each with $\gamma = 2$. In (c), the picture has been split again, into four pentagonal vertices, each with $\gamma = 1$.

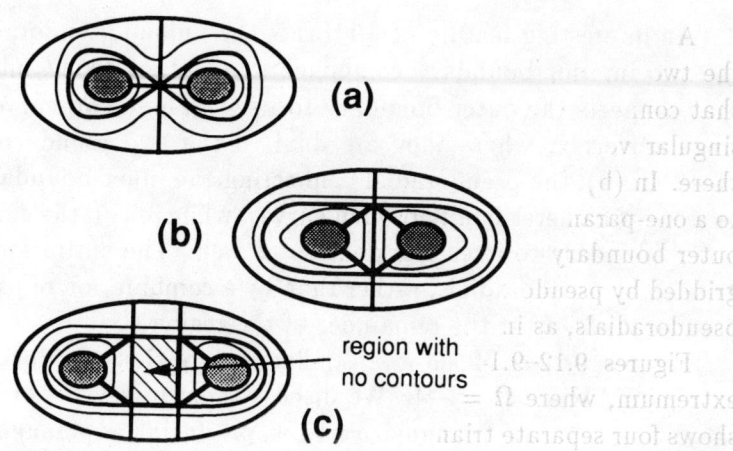

region with
no contours

FIG. 9.10.   *Pseudocontours at a saddle.*

Figure 9.10 shows the consequences of these choices for the pseudocontours, which are just the curves constructed by stringing together isoparameter segments parametrically parallel to the boundary curves. Figure 9.10(a) shows the contours we would expect near a typical saddle point. In (b), the saddle point has been extended into a saddle ridge along the edge connecting the two hexagonal vertices. This ridge has now become an asymptote for pseudocontours. In (c), the saddle ridge curve has been further extended into a saddle region, the crosshatched rectangle. The boundaries of this region are asymptotes for pseudocontours. No pseudocontours at all can be drawn inside the region; any isoparameter curve in the saddle region extends to a curve that crosses the pseudocontours already drawn.

It is also instructive to consider the complementary family of curves, constructed by stringing together segments parametrically normal to the boundary. Such curves will be called *pseudoradials*. Figure 9.11 shows representative pseudoradial curves for the three constructions of Figure 9.9.

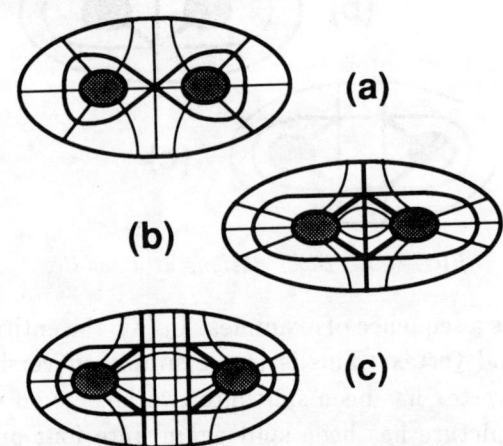

FIG. 9.11. *Pseudoradials at a saddle.*

An interesting feature of 9.11(a) is the unique pseudoradial that connects the two interior boundary components, and the second unique pseudoradial that connects the outer boundary to itself. These two curves intersect at the singular vertex, where they are dual to the two pseudocontours that meet there. In (b), the pseudoradial connecting the inner boundaries has expanded to a one-parameter family of such curves, while in (c), the curve connecting the outer boundary to itself has done so as well. The entire saddle region is now gridded by pseudoradials, rather than by a combination of pseudocontours and pseudoradials, as in the remainder of the region.

Figures 9.12–9.14 show a similar progression of block designs near an extremum, where $\Omega = -4$. We discuss them in reverse order. Thus, 9.12(c) shows four separate triangular vertices, producing a rectangular region gridded entirely by pseudoradials, in which no pseudocontours exist. In (b), this region has been collapsed to a ridge line, and in (a) the line collapses to a point extremum.

The tradeoff, however, is that in (a) and (b) formerly rectangular patches have now collapsed along one edge apiece to become geometrically triangular, although still smooth and parametrically rectangular. Parametric singularities of this sort do not affect at all the construction of pseudocontours and pseudoradials by evaluation of isoparameter curves. Smooth pseudocontours

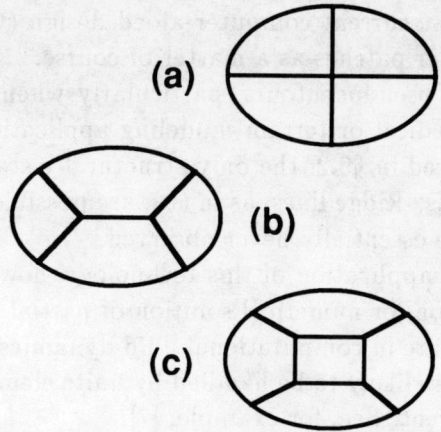

FIG. 9.12. *Blockings at an extremum.*

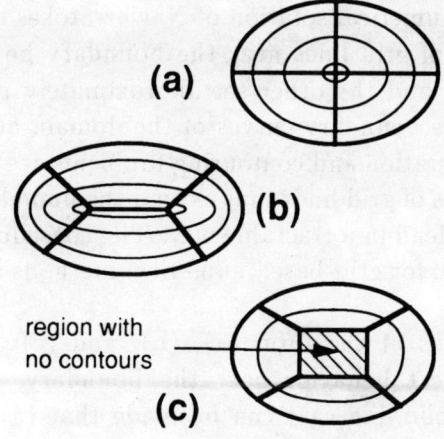

region with
no contours

FIG. 9.13. *Pseudocontours at an extremum.*

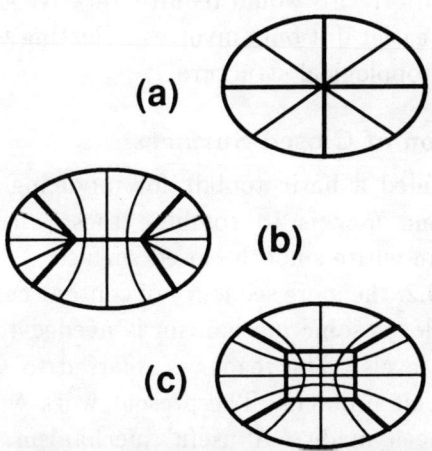

FIG. 9.14. *Pseudoradials at an extremum.*

extracted by plane cuts of the geometrically smooth surface should similarly be unaffected, since most current computer aided design (CAD) systems handle degenerate rectangular patches as a matter of course.

For constructing pseudocontours, particularly when dealing with natural phenomena, as in medical or terrain modeling applications, alternative (a) is preferred. As remarked in §9.2, the only structurally stable situation calls for isolated critical points. Ridge lines, as in (b), are possible but more rare, while plateaus as in (c) are essentially never observed.

Another possible application of this technology, however, lies in finite difference grid generation for numerical solution of partial differential equations. Typical applications are in computational fluid dynamics, since mechanical engineering problems are likely to be handled by finite element techniques, where the criteria are different. See, for example, [7].

In the grid generation context, the machinery described in §9.2 corresponds closely to the notion of a *multiblock grid design*, in which a topologically rectangular grid is defined in each face of a tessellation of the computational domain. As remarked in [22], numerical solution of Navier-Stokes equations essentially requires that one set of grid lines near the boundary be approximately parallel to the boundary, and the other set approximately perpendicular to the boundary. Thus, if the boundary curves of the domain are regarded as input contours, the grid generation and contouring problems are essentially identical: how to develop families of grid lines that extend the boundary curves smoothly into the interior and blend in a tractable way. The consequences of exceptional vertices at which $\gamma \neq 0$ for grid-based numerical methods are also discussed at length in [22].

In practice, designs of the form (a), (b), and (c) are all useful, since they all provide correct behavior near the boundary, and all are in fact used. However, the following case can be made that (a) might be preferred in some circumstances. It is computationally convenient to have grid lines approximately aligned with fluid flow everywhere. Since streamlines are, of course, not known a priori, this would require iterative grid adjustment. That, in turn, will be much easier if it only involves adjusting the geometry of a grid, rather than its basic topological structure.

9.5. Reconstruction of Closed Surfaces

Section 9.4 has provided a basic toolkit for modeling features of a smooth surface. The problem now is to combine these structures to produce a parameterization of an entire smooth closed surface.

As remarked in §9.2, the mere sequence of contour curves does not uniquely define a surface topology; some mechanism is needed to match contour components on adjacent levels. The reader is referred to the references cited in §9.1 for insight into this problem. The present work will simply assume that the decisions have been made. A useful mechanism for encoding them is

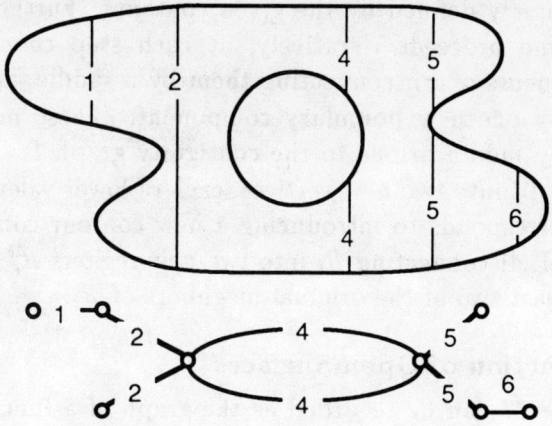

FIG. 9.15. *Contiguity graph for a closed surface.*

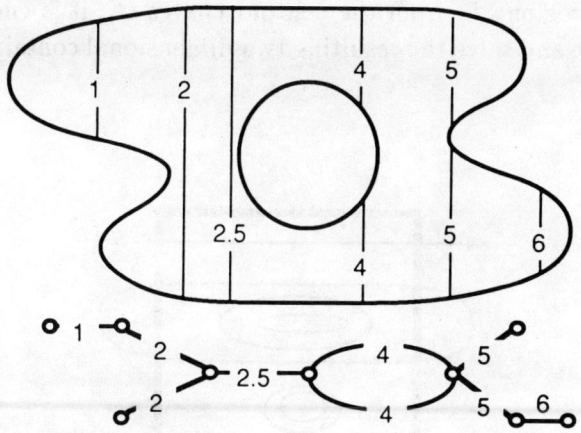

FIG. 9.16.   *Subdividing a multiple saddle.*

provided by the contour contiguity graph construction of [16], which we now describe.

Let M be a surface, and $\{C_j\}_{j=1}^N$ be a set of contour curves on it. Then M is separated by the curves C_j into a set of connected regions, denoted R_i, which correspond to the vertices of a graph Γ. The edges of Γ correspond to the set of curves C_j. An edge connects two vertices precisely when the curve separates the two regions. See Fig. 9.15 for an example.

Parameterizing M now reduces to finding a block design for each region R_i. Note that each region is classified by its valence in Γ, that is, by the number of other regions to which it is linked. A region of valence 1 is an extremum, and can be blocked as in Fig. 9.12. A region of valence 2 is topologically an open cylinder, and can be blocked trivially with one rectangular patch, or more if desired.

A region of valence $k + 2$ is a k-fold saddle. That is, it can be assumed to contain k distinct saddle points. However, the location of these saddle points is not uniquely defined by the given contours. Further decisions need to be made. One proceeds iteratively, at each step choosing two of the boundary components of R_i, connecting them by a saddle, and replacing two old boundaries by one new boundary component. These manipulations are easily recorded by modifications to the contiguity graph Γ. The vertex R_i of valence $k+2$ is split into two new vertices, each of lower valence. As shown in Fig. 9.16, this corresponds to introducing a new contour component at some intermediate level, disconnecting R_i into two new regions R_i^+ and R_i^-, each of which abuts at least two of the original nieghbors of R_i.

9.6. Reconstruction of Open Surfaces

When the surface M can be regarded as the graph of a function $z = f(x, y)$, for example in terrain modeling applications, additional considerations arise, which we now briefly discuss.

To begin with, the ambiguity illustrated by Figs. 9.3 and 9.4 cannot occur. To identify the regions R_i from the contour curves C_j, it is enough to project into the xy plane and solve the resulting two-dimensional containment problem.

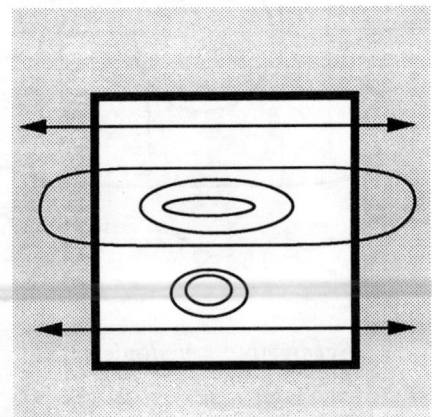

FIG. 9.17. *Contours and a clipping box.*

This works fine as long as all the projected contours actually separate the plane, for example if all are closed contours, or open contours that are known to stretch to infinity. Often, however, the input curves C_j are merely the fragments of such separating contours that happen to lie in some bounded domain to which f has been restricted. Figure 9.17 shows a typical set of contours in the plane, and the fragments visible in the unshaded region inside a rectangular clipping box. Figure 9.18 illustrates an alternative reconstruction of the global contour structure from the same fragments. Resolving these ambiguous cases is the main topological difficulty of explicit surface reconstruction. A compensating advantage, however, is that some of these decisions can be hedged by locating the implied singularities beyond the limits of the domain of interest, and then never explicitly modeling the details.

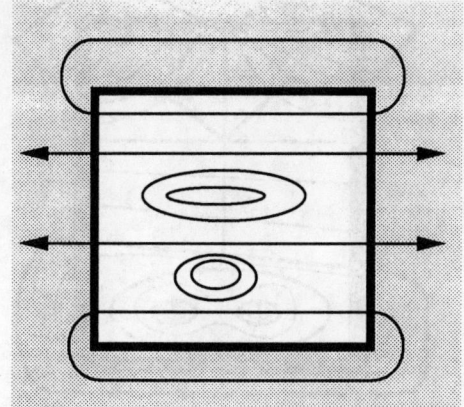

FIG. 9.18.   *An alternative reconstruction.*

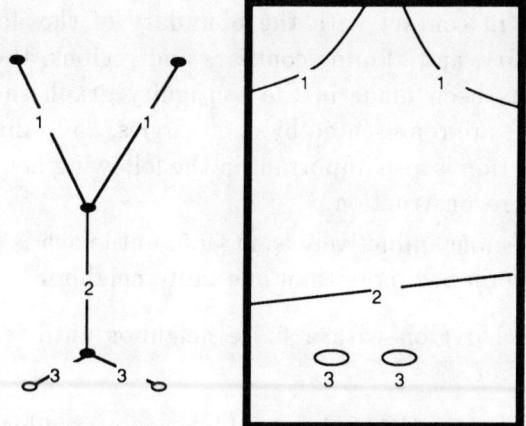

FIG. 9.19.   *Contiguity graph for contours over a bounded domain.*

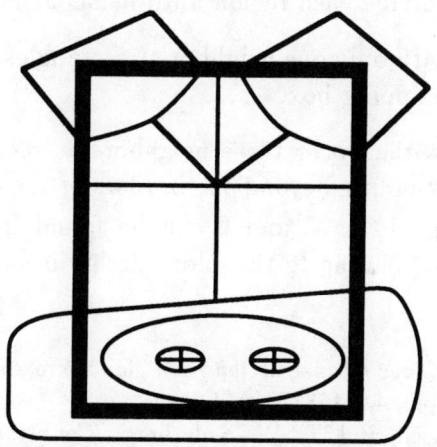

FIG. 9.20.   *Blocking extends beyond clipping box.*

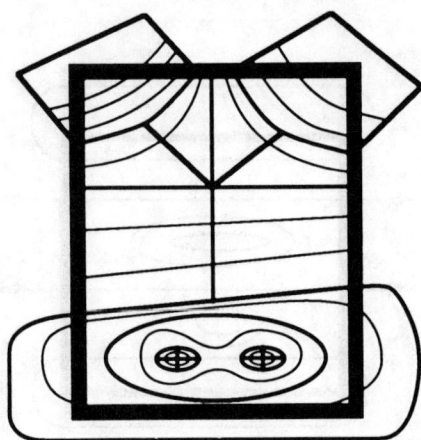

FIG. 9.21. *Pseudocontours extend beyond box.*

Evidently, there is a fundamental distinction between "infinite" contours and regions, those in contact with the boundary of the domain, which can give rise to ambiguity, and "finite" contours and regions, those which do not. This distinction has been made in the contiguity graph shown in Fig. 9.19, where finite vertices are represented by open circles, and infinite ones by black circles. This distinction is also important in the following heuristic prescription for explicit surface reconstruction.

1. Treat finite regions inductively as in §9.5, until each is fully blocked, and no infinite region has more than one finite neighbor.

2. Subdivide each region with a finite neighbor until it has at most one infinite neighbor.

3. Block each region with a finite neighbor as an annulus.

4. At this point, the remaining regions are infinite, with only infinite neighbors. Subdivide each region until it has at most three neighbors.

5. Block regions with only one neighbor as rectangles extending beyond the borders of the clipping box.

6. Block regions with two or three neighbors as rectangles and hexagons, respectively, extending beyond the borders of the clipping box.

Examples of steps 1, 3, 5, and 6 can be found in Fig. 9.20. A set of pseudocontours corresponding to this block design is shown in Fig. 9.21.

References

[1] J.-D. Boissonat, *Shape reconstruction from planar cross sections*, Comput. Vision Graph. Image Proc., 44 (1988), pp. 1–29.

[2] W. H. Christiansen and T. W. Sederberg, *Conversion of complex contour line definitions into polygonal element mosaics*, Comput. Graphics, 12 (1978), pp. 187–192.

[3] L. J. Dickson, *Geometry Class*, unpublished notes, Boeing Commercial Airplane Co., Seattle, WA, 1978.

[4] P. A. Dowd, *Algorithms for three-dimensional interpolation between planar slices*, in Fundamental Algorithms for Computer Graphics, R. A. Earnshaw, ed., NATO ASI Series F, vol. 17, Springer-Verlag, New York, 1985.

[5] O. D. Faugeras, M. Hebert, P. Mussi, and J. D. Boissonat, *Polyhedral approximation of 3-D objects without holes*, Comput. Vision Graph. Image Proc., 25 (1984), pp. 169–183.

[6] H. Fuchs, Z. Kedem, and S. P. Uselton, *Optimal surface reconstruction from planar contours*, Comm. ACM, 20 (1977), pp. 693–702.

[7] T. S. Hlaban, *Quadrilateral tessellation*, preprint, GE Medical Systems, Milwaukee, WI, November 1989.

[8] A. K. Jones, *Nonrectangular surface patches with curvature continuity*, Comput. Aided Des., 20 (1988), pp. 325–335.

[9] ——, *Grid singularities in the plane*, Tech. report ETA-TR-82, Boeing Computer Services, Seattle, WA, 1988.

[10] ——, *Mathematical tools for 3D multiblock grid design*, Tech. report SCA-TR-121, Boeing Computer Services, Seattle, WA, 1989.

[11] E. Keppel, *Approximating complex surfaces by triangulating of contour lines*, IBM J. Res. Develop., 19 (1975), pp. 2–11.

[12] C. Lin, *Shape and texture from serial contours*, Mathematical Geol., 15 (1983), pp. 617–632.

[13] W.-C. Lin, S.-Y. Chen and C.-T. Chen, *A new surface interpolation technique for reconstructing 3D objects from serial cross-sections*, Comput. Vision Graph. Image Proc., 48 (1989), pp. 124–143.

[14] J. Milnor, *Morse Theory*, Princeton University Press, Princeton, NJ, 1969.

[15] M. Morse, *A positive, lower semi-continuous non-degenerate function on a metric space*, Fund. Math., 35 (1948), pp. 47–78.

[16] S. P. Morse, *Concepts of use in contour map processing*, Comm. ACM, 12 (1969), pp. 147–152.

[17] J. Peters, *Smooth mesh interpolation with cubic patches*, Comput. Aided Des., 22 (1990), pp. 109–120.

[18] T. Poston and I. N. Stewart, *Taylor Expansions and Catastrophes*, Pitman, London, 1976.

[19] A. Roberts, *Automatic topology generation and generalised B-spline mapping*, in Numerical Grid Generation, J. F. Thompson, ed., Elsevier, Amsterdam, 1982.

[20] R. F. Sarraga, *Computer modeling of surfaces with arbitrary shapes*, IEEE Comput. Graph. Appl., 10 (1990), pp. 67–77.

[21] M. Shantz, *Surface definition for branching contour-defined objects*, Comput. Graphics, 15 (1981), pp. 242–270.

[22] J. F. Thompson, Z. U. A. Warsi, and C. W. Mastin, *Numerical Grid Generation*, North-Holland, New York, 1985.

[23] Y. F. Wang and J. K. Aggarwal, *Surface reconstruction and representation of 3D scenes*, Pattern Recognition, 19 (1986), pp. 197–207.

Conditions and Constructions for GC^1 and GC^2 Continuity Between Adjacent Integral Bézier Surface Patches

Peter Wassum

10.1. Introduction

Collections of surface patches are usually applied for modeling free-form surfaces indicating the importance of considering continuity conditions for adjacent patches. Attention will be focused on the case of two adjacent surface patches joining smoothly along a common boundary curve. The considered surfaces may be assumed to be integral tensor-product or triangular Bézier patches which are extensively used for surface shape controlling in computer aided geometric design (CAGD). Avoiding unnecessary restrictions due to parameterization, geometric continuity (GC^h-continuity) is widely regarded as the suitable method to connect neighboring surface patches.

Geometric C^1- and C^2-continuity conditions concerning adjacent Bézier patches of tensor product or triangular type have been reflected in numerous papers ([1], [3]–[6], [11]–[16], [17]–[21]). The linear equation system of Farin's GC^1-construction [10] involves Bézier control points of the common edge and of the first neighboring column on each side of the boundary curve as well as scalar form parameters.

Liu and Hoschek [17] determined necessary and sufficient vector-valued conditions for geometric C^1-continuity of two adjacent Bézier patches. Their solution, like the one given by Farin, is composed of a linear equation system in different vectors between corresponding control points and unknown shape parameters.

DeRose [9] derived necessary and sufficient scalar GC^1-constraints on the position of Bézier points and showed their linear independence in the general case. These conditions allow control of the geometric C^1-continuous joint of two given Bézier patches.

Degen's paper [8] concerning necessary and sufficient vector-valued geometric C^1- and C^2-constraints yields explicit representations for involved Bézier control points of the first and second column next to the common boundary. His method requires information on both the common edge and the tangent plane in each point of the common boundary in an appropriate way.

As an extension of Liu's and Hoschek's results, this paper introduces necessary and sufficient geometric C^2-constraints covering all four combinations of tensor-product and triangular integral Bézier patches. A system of necessary and sufficient geometric C^2-constraints on the positions of control points is investigated from the viewpoint of the independence of these constraints in the general case and provides the possibility of controlling the second-order geometric continuous joint of the two given surface patches.

In developing special sufficient geometric C^1- and C^2-constraints for practical purposes, we will confine ourselves to the case of adjacent tensor-product patches. With Bézier points of the common boundary and first and second interior column of the control mesh of one patch assumed as being given, these constraints, including a suitable number of shape parameters, can be viewed as construction methods for the unknown control points of the neighboring patch and will be investigated concerning their geometric properties.

10.2. Fundamentals

10.2.1. Notation.

10.2.1.1. *Tensor-product and triangular integral Bézier patches.* Expanded in terms of Bernstein polynomials

$$B_k^n(u) = \begin{pmatrix} n \\ k \end{pmatrix} u^k(1-u)^{n-k}, \qquad k = 0(1)n, \quad u \in [0,1],$$

involved integral tensor-product Bézier patches $\mathbf{X}$ and $\mathbf{Y}$ of degree (n, m) and (n^*, m^*), respectively, have the representation

$$\mathbf{X}(r,s) = \sum_{i=0}^{n} \sum_{j=0}^{m} \mathbf{E}_{ij} B_i^n(r) B_j^m(s), \qquad \mathbf{E}_{ij} \in \Re^3, \qquad (r,s) \in [0,1] \times [0,1],$$

(10.1)

$$\mathbf{Y}(u,v) = \sum_{i=0}^{n^*} \sum_{j=0}^{m^*} \mathbf{F}_{ij} B_i^{n^*}(u) B_j^{m^*}(v), \qquad \mathbf{F}_{ij} \in \Re^3, \quad (u,v) \in [0,1] \times [0,1],$$

with Bézier control points $\{\mathbf{E}_{ij}\}$, $\{\mathbf{F}_{ij}\}$ describing the corresponding control meshes. The common edge attached to patch $\mathbf{X}$ will be given by

$$s = 0, \quad r \in [0,1] : \mathbf{X}(r,s)_{|\mathbf{B}} = \mathbf{X}(0,s)$$

and to patch $\mathbf{Y}$ by

$$u = 1, \quad v \in [0,1] : \mathbf{Y}(u,v)_{|\mathbf{B}} = \mathbf{Y}(1,v).$$

Referring to generalized Bernstein polynomials

$$B_{ijk}^m(u,v,w) = \frac{m!}{i!\,j!\,k!} u^i v^j w^k$$

$$u + v + w = 1, \quad u, v, w \geq 0, \quad i + j + k = m,$$

considering integral triangular Bézier patches $\mathbf{X}$ and $\mathbf{Y}$ of degree m and m^*, respectively, are represented by

$$\mathbf{X}(r,s,t) = \sum_{i+j+k=m} \mathbf{E}_{ijk} B_{ijk}^m(r,s,t), \quad \mathbf{E}_{ijk} \in \Re^3, \quad r+s+t=1, \; r,s,t, \geq 0,$$

$$\mathbf{Y}(u,v,w) = \sum_{i+j+k=m^*} \mathbf{F}_{ijk} B_{ijk}^{m^*}(u,v,w), \quad \mathbf{F}_{ijk} \in \Re^3,$$

$$u + v + w = 1, \; u, v, w \geq 0,$$

(10.2)

with Bézier control points $\{\mathbf{E}_{ijk}\}$, $\{\mathbf{F}_{ijk}\}$ forming the associated control meshes. The common edge linked to patch $\mathbf{X}$ will be described by

$$s = 0, \quad r \in [0,1] : \mathbf{X}(r,s,t)_{|\mathbf{B}} = \mathbf{X}(0,s,1-s)$$

and to patch $\mathbf{Y}$ by

$$u = 0, \quad v \in [0,1] : \mathbf{Y}(u,v,w)_{|\mathbf{B}} = \mathbf{Y}(1,v,1-v).$$

Varying along the same integral curve, parameters s and v are connected by a linear transformation and boundary values of this transformation identify s and v along the common boundary [8]: $s_{|\mathbf{B}} = v, v \in [0,1]$. The notation for the partial derivatives of the tensor-product patches $\mathbf{X}$ and $\mathbf{Y}$ along the common edge is now introduced, corresponding to an analogous notation for the directional derivatives of triangular patches [11]. Cross derivatives of first order and mixed derivatives of second order involving control points of the common edge and the first interior column of each control mesh next to the boundary are denoted by

$$\mathbf{D}_r\mathbf{X}(v) := \frac{\delta}{\delta r}\mathbf{X}(r,s)_{|\mathbf{B}} = n \sum_{j=0}^{m} \mathbf{a}_j^{(1)} B_j^m(v) =: \sum_{j=0}^{m} \overline{\mathbf{a}}_j B_j^m(v),$$

$$\mathbf{D}_u\mathbf{Y}(v) := \frac{\delta}{\delta u}\mathbf{Y}(u,v)_{|\mathbf{B}} = n^* \sum_{j=0}^{m^*} \mathbf{c}_j^{(1)} B_j^{m^*}(v) =: \sum_{j=0}^{m^*} \overline{\mathbf{c}}_j B_j^{m^*}(v),$$

$$\mathbf{D}_{rv}\mathbf{X}(v) := \frac{\delta^2}{\delta r\, \delta s}\mathbf{X}(r,s)_{|\mathbf{B}} = nm \sum_{j=0}^{m-1} \left(\mathbf{a}_{j+1}^{(1)} - \mathbf{a}_j^{(1)} \right) B_j^{m-1}(v)$$

(10.3)
$$=: \sum_{j=0}^{m-1} \overline{\mathbf{q}}_j B_j^{m-1}(v),$$

$$\mathbf{D}_{uv}\mathbf{Y}(v) := \frac{\delta^2}{\delta u\, \delta v}\mathbf{Y}(u,v)_{|\mathbf{B}} = n^*m^* \sum_{j=0}^{m^*-1} \left(\mathbf{c}_{j+1}^{(1)} - \mathbf{c}_j^{(1)} \right) B_j^{m^*-1}(v)$$

$$=: \sum_{j=0}^{m^*-1} \overline{\mathbf{r}}_j B_j^{m^*-1}(v).$$

Each of the cross derivatives of second order given by the expressions

$$\mathbf{D}_{rr}\mathbf{X}(v) := \frac{\delta^2}{\delta r^2}\mathbf{X}(r,s)_{|\mathbf{B}} = n(n-1)\sum_{j=0}^{m}\left(\mathbf{a}_j^{(2)} - \mathbf{a}_j^{(1)}\right)B_j^m(v) =: \sum_{j=0}^{m}\overline{\mathbf{s}}_j B_j^m(v),$$

$$\mathbf{D}_{uu}\mathbf{Y}(v) := \frac{\delta^2}{\delta u^2}\mathbf{Y}(u,v)_{|\mathbf{B}} = n^*(n^*-1)\sum_{j=0}^{m^*}\left(\mathbf{c}_j^{(1)} - \mathbf{c}_j^{(2)}\right)B_j^{m^*}(v)$$

$$=: \sum_{j=0}^{m^*}\overline{\mathbf{t}}_j B_j^{m^*}(v),$$

(10.4)

contains the next neighboring column of the adjoined control mesh.

The introduced $\mathbf{a}_j^{(k)}, \mathbf{c}_j^{(k)}$ denote difference vectors combining control points

$$\mathbf{a}_j^{(k)} := \mathbf{E}_{k,j} - \mathbf{E}_{k-1,j}, \quad \mathbf{c}_j^{(k)} := \mathbf{F}_{n^*-k+1,j} - \mathbf{F}_{n^*-k,j} \qquad (k=1,2).$$

10.2.1.2. *Common edge.* With $\{\mathbf{E}_j\}$ as associated set of control points, the notation

(10.5) $$\mathbf{B}(v) = \sum_{j=0}^{\overline{p}}\mathbf{E}_j B_j^{\overline{p}}(v), \quad \mathbf{E}_j \in \Re^3, \quad v \in [0,1]$$

is introduced to represent the common boundary curve of degree $\overline{p}$.

Derivatives of first and second order of the common edge $\mathbf{B}(v)$ are described by

$$\mathbf{D}_v\mathbf{X}(v) \quad := \frac{\partial}{\partial v}\mathbf{B}(v) = \overline{p}\sum_{j=0}^{\overline{p}-1}\mathbf{d}_j B_j^{\overline{p}-1}(v) =: \sum_{j=0}^{\overline{p}-1}\overline{\mathbf{d}}_j B_j^{\overline{p}-1}(v),$$

$$\mathbf{D}_{vv}\mathbf{X}(v) := \frac{\partial^2}{\partial v^2}\mathbf{B}(v) = \overline{p}(\overline{p}-1)\sum_{j=0}^{\overline{p}-2}(\mathbf{d}_{j+1} - \mathbf{d}_j)B_j^{\overline{p}-2}(v) =: \sum_{j=0}^{\overline{p}-2}\overline{\mathbf{e}}_j B_j^{\overline{p}-2}(v)$$

(10.6)

with

$$\mathbf{d}_j := \mathbf{E}_{j+1} - \mathbf{E}_j.$$

Using integral representations, directional derivatives of tensor-product and triangular Bézier patches as well as derivatives of the common boundary are vector-valued polynomials depending on the parameter v.

Composing two adjacent tensor-product and triangular Bézier patches, four combination variants are imaginable.

(i) Combination "tensor-product patch – tensor-product patch," common boundary curve $\mathbf{Y}(1,v) = \mathbf{X}(0,v)$. See Fig. 10.1.

(ii) Combination "triangular patch – triangular patch," common boundary curve $\mathbf{Y}(0,v,1-v) = \mathbf{X}(0,v,1-v)$ (implying opposite orientation of the common edge). See Fig. 10.2.

(iii) Combination "tensor-product patch – triangular patch," common boundary curve $\mathbf{Y}(1,v) = \mathbf{X}(0,v,1-v)$. See Fig. 10.3.

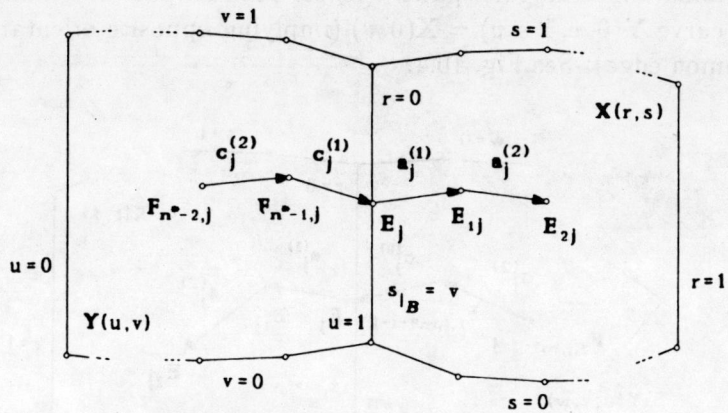

FIG. 10.1 *Adjacent patches of type "tensor-product – tensor-product."*

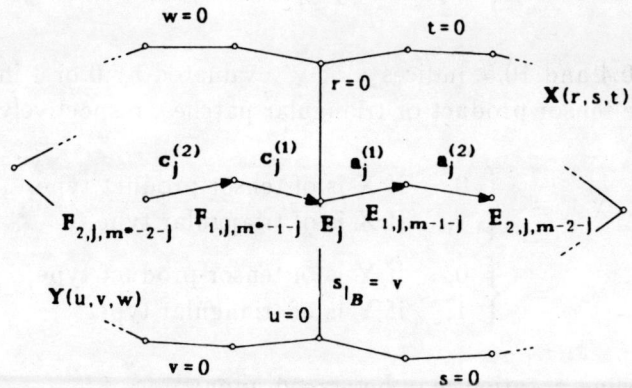

FIG. 10.2 *Adjacent patches of type "triangular – triangular."*

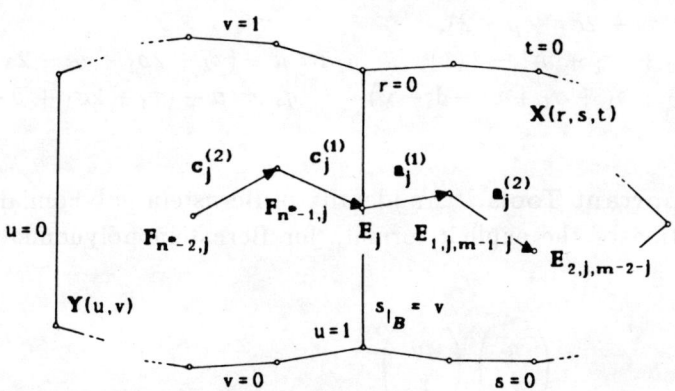

FIG. 10.3 *Adjacent patches of type "tensor-product – triangular."*

(iv) Combination "triangular patch – tensor-product patch," common boundary curve $\mathbf{Y}(0, v, 1 - v) = \mathbf{X}(0, v)$ (implying opposite orientation of the common edge). See Fig. 10.4.

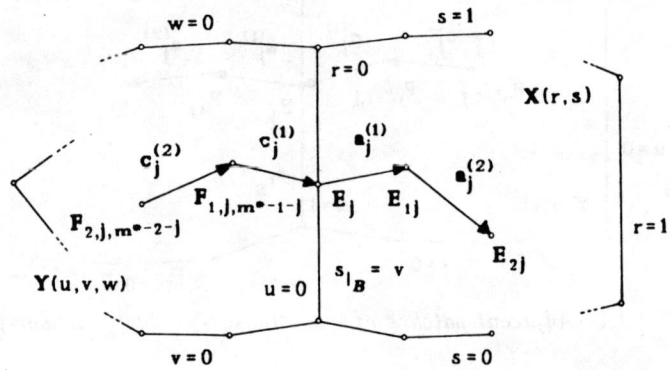

FIG. 10.4 *Adjacent patches of type "triangular – tensor-product."*

Covering §§10.4 and 10.5, indices χ or χ^* evaluated by 0 or 1 indicate that $\mathbf{X}$ or $\mathbf{Y}$ describe tensor-product or triangular patches, respectively

(10.7)
$$\chi = \begin{cases} 0, & \text{if } \mathbf{X} \text{ is of tensor-product type,} \\ 1, & \text{if } \mathbf{X} \text{ is of triangular type,} \end{cases}$$
$$\chi^* = \begin{cases} 0, & \text{if } \mathbf{Y} \text{ is of tensor-product type,} \\ 1, & \text{if } \mathbf{Y} \text{ is of triangular type,} \end{cases}$$

and the following notation may be agreed upon:

$t := \max(m + \bar{p}, m^* + \bar{p}, m^* + m),$
$k_1 := t - (m + \bar{p}), \quad k_2 := t - (m^* + \bar{p}), \quad k_3 := t - (m^* + m),$
$\mu := \max(\tau_2 + 2\tau_1 + m^* - 2\chi^*, \tau_2 + 2\rho_1 + m - 2\chi, \tau_2 + \rho_1 + \sigma_1 + m - 1 - \chi,$
$\qquad \tau_2 + 2\sigma_1 + \bar{p} - 2),$
$g_1 := \mu - (\tau_2 + 2\tau_1 + m^* - 2\chi^*), \qquad g_2 := \mu - (\tau_2 + 2\rho_1 + m - 2\chi),$
$g_3 := \mu - (\tau_2 + \rho_1 + \sigma_1 + m - 1 - \chi), \qquad g_4 := \mu - (\tau_2 + 2\sigma_1 + \bar{p} - 2).$
(10.8)

10.2.2. Important Tools.

An identity of Bernstein polynomials obtained by manipulation of the explicit formula for Bernstein polynomials is represented by

$$(10.9) \quad B_i^n(v) B_j^m(v) = \frac{\binom{n}{i}\binom{m}{j}}{\binom{n+m}{i+j}} B_{i+j}^{n+m}(v), \quad i = 0(1)n, \quad j = 0(1)m.$$

Based on generalized degree elevation, a Bézier curve of degree p can also be expressed as a Bézier curve of degree $p + s$:

$$\sum_{j=0}^{p} \mathbf{W}_j B_j^p(v) = \sum_{j=0}^{p+s} [\mathbf{W}_j]_s B_j^{p+s}(v), \quad \mathbf{W}_j \in \Re^3$$

(10.10)

$$\text{with } [\mathbf{W}_j]_s := \sum_{k=j-s}^{j} \frac{\binom{p}{k}\binom{s}{j-k}}{\binom{p+s}{j}} \mathbf{W}_k, \qquad (j = 0(1)p + s).$$

10.3. Conditions of Contact of Order

Introducing geometric C^h-continuity ($h = 1,2$) conditions we will refer to osculating conditions which are well known in differential geometry as conditions of contact of order h of two surfaces. Throughout the following considerations, neighboring tensor-product and triangular Bézier surfaces are assumed to be regular in each point $\mathbf{B}(\overline{v}), \overline{v} \in [0,1]$ of the common boundary

$$\mathbf{D}_r\mathbf{X}(\overline{v}) \times \mathbf{D}_v\mathbf{X}(\overline{v}) \neq 0, \quad \mathbf{D}_u\mathbf{Y}(\overline{v}) \times \mathbf{D}_v\mathbf{X}(\overline{v}) \neq 0, \quad \text{for all } \overline{v} \in [0,1],$$

implying linear independence of the derivative vectors $\mathbf{D}_r\mathbf{X}(\overline{v}), \mathbf{D}_v\mathbf{X}(\overline{v})$ as well as of $\mathbf{D}_u\mathbf{Y}(\overline{v}), \mathbf{D}_v\mathbf{X}(\overline{v})$ in each point $\mathbf{B}(\overline{v})$.

Two adjacent surfaces join geometric C^1-continuously at a common point $\mathbf{P} = \mathbf{B}(\overline{v})$ if the following conditions hold at $\mathbf{P}$ (arbitrary form coefficients with $r_{ij}, s_{ij}(i,j = 0,1)$ [7])

$$(10.11) \qquad h = 1: \begin{pmatrix} \mathbf{D}_u\mathbf{Y}(\overline{v}) \\ \mathbf{D}_v\mathbf{Y}(\overline{v}) \end{pmatrix} = \begin{pmatrix} r_{10} & s_{10} \\ r_{01} & s_{01} \end{pmatrix} \cdot \begin{pmatrix} \mathbf{D}_r\mathbf{X}(\overline{v}) \\ \mathbf{D}_v\mathbf{X}(\overline{v}) \end{pmatrix}.$$

Two adjacent surfaces have a second-order geometric continuous joint at a common point $\mathbf{P} = \mathbf{B}(\overline{v})$ if the following conditions hold at $\mathbf{P}$ (with arbitrary form coefficients $r_{ij}, s_{ij}(i,j = 0,1,2)$)

$$h = 2: \begin{pmatrix} \mathbf{D}_{uu}\mathbf{Y}(\overline{v}) \\ \mathbf{D}_{uv}\mathbf{Y}(\overline{v}) \\ \mathbf{D}_{vv}\mathbf{Y}(\overline{v}) \end{pmatrix} = \begin{pmatrix} r_{20} & s_{20} \\ r_{11} & s_{11} \\ r_{02} & s_{02} \end{pmatrix} \cdot \begin{pmatrix} \mathbf{D}_r\mathbf{X}(\overline{v}) \\ \mathbf{D}_v\mathbf{X}(\overline{v}) \end{pmatrix}$$

$$+ \begin{pmatrix} r_{10}^2 & 2r_{10}s_{10} & s_{10}^2 \\ r_{10}s_{01} & r_{10}s_{01} + r_{01}s_{10} & s_{10}s_{01} \\ r_{01}^2 & 2r_{01}s_{01} & s_{01}^2 \end{pmatrix} \cdot \begin{pmatrix} \mathbf{D}_{rr}\mathbf{X}(\overline{v}) \\ \mathbf{D}_{rv}\mathbf{X}(\overline{v}) \\ \mathbf{D}_{vv}\mathbf{X}(\overline{v}) \end{pmatrix}$$

(and conditions attached to $h = 1$).

(10.12)

The GC^1- and GC^2-continuity constraints (10.11), (10.12) correspond to geometric interpretations. Surfaces with geometric C^1-continuity in a common point $\mathbf{P}$ have a common tangent plane in $\mathbf{P}$; geometric C^2-continuous surfaces in a common point $\mathbf{P}$ are tangent plane and Dupin's indicatrix continuous in $\mathbf{P}$.

The following considerations will treat the geometric C^1- and C^2-continuous joint of two adjacent patches $\mathbf{X}$ and $\mathbf{Y}$ along their common boundary $\mathbf{B}(v)$ (positional continuity). Hence shape coefficients $r_{ij}, s_{ij}(i, j, = 0, 1, 2)$ represent functions of the parameter v in general. The following considerations will show that the system of resulting geometric continuity conditions to be observed consists of h constraints, one condition to each h ($h = 1, 2$).

Regarding derivative vectors of first and second order along the direction of the common boundary $\mathbf{B}(v)$, we obtain the equations

$$\mathbf{D}_v\mathbf{Y}(v) = \mathbf{D}_v\mathbf{X}(v),$$
$$\mathbf{D}_{vv}\mathbf{Y}(v) = \mathbf{D}_{vv}\mathbf{X}(v),$$

which yield the following identities concerning the involved shape parameter functions

$$r_{01}(v) = r_{02}(v) = s_{02}(v) \equiv 0, \qquad s_{01}(v) \equiv 1.$$

The geometric C^1-condition (condition of contact of order 1) describing tangent plane continuity of the adjacent surface patches along $\mathbf{B}(v)$ therefore reduces to

(10.13) $\qquad \mathbf{D}_u\mathbf{Y}(v) = r_{10}(v)\mathbf{D}_r\mathbf{X}(v) + s_{10}(v)\mathbf{D}_v\mathbf{X}(v),$

with scalar functions $r_{10}(v)$, $s_{10}(v)$ (denoted as coefficient functions) and involves control points of the common edge and the first interior column of each control mesh next to the boundary $\mathbf{B}(v)$.

Extending to geometric C^2-continuity of neighboring surfaces, we have to take into account conditions attached to mixed derivative $\mathbf{D}_{uv}\mathbf{Y}(v)$ and cross derivative of second order $\mathbf{D}_{uu}\mathbf{Y}(v)$ being formed as linear combinations of derivative vectors of first and second order linked to patch $\mathbf{X}$.

Please note that the condition for the mixed derivative vector

(10.14)
$$\mathbf{D}_{uv}\mathbf{Y}(v) = \frac{d[r_{10}(v)]}{dv}\mathbf{D}_r\mathbf{X}(v) + \frac{d[s_{10}(v)]}{dv}\mathbf{D}_v\mathbf{X}(v)$$
$$+ r_{10}(v)\mathbf{D}_{rv}\mathbf{X}(v) + s_{10}(v)\mathbf{D}_{vv}\mathbf{X}(v)$$

is easily obtained by differentiating the geometric C^1-condition (10.13) with respect to v and is automatically satisfied if (10.13) holds. Therefore the condition on the cross derivative of second order

(10.15)
$$\mathbf{D}_{uu}\mathbf{Y}(v) = r_{20}(v)\mathbf{D}_r\mathbf{X}(v) + s_{20}(v)\mathbf{D}_v\mathbf{X}(v) + [r_{10}(v)]^2\mathbf{D}_{rr}\mathbf{X}(v)$$
$$+ 2r_{10}(v)s_{10}(v)\mathbf{D}_{rv}\mathbf{X}(v) + [s_{10}(v)]^2\mathbf{D}_{vv}\mathbf{X}(v)$$

determines Dupin's indicatrix continuity of the adjacent surface patches along $\mathbf{B}(v)$ and will be denoted as geometric C^2-condition (condition of contact of order 2).

10.4. Geometric C^1-Continuity

The geometric C^1-condition (10.13) describes the derivative vector $\mathbf{D}_u\mathbf{Y}(v)$ as being coplanar to the linear independent derivative vectors $\mathbf{D}_r\mathbf{X}(v)$ and $\mathbf{D}_v\mathbf{X}(v)$ which span the tangent plane in each point of the boundary $\mathbf{B}(v)$. See Fig. 10.5. According to the well-known determinant characterization of tangent plane continuity [17], two adjacent patches join GC^1-continuously in every point of the common boundary if and only if

$$(10.16) \qquad \Phi(v) = \det\{\mathbf{D}_u\mathbf{Y}(v), \mathbf{D}_r\mathbf{X}(v), \mathbf{D}_v\mathbf{X}(v)\} = 0.$$

See Fig. 10.5.

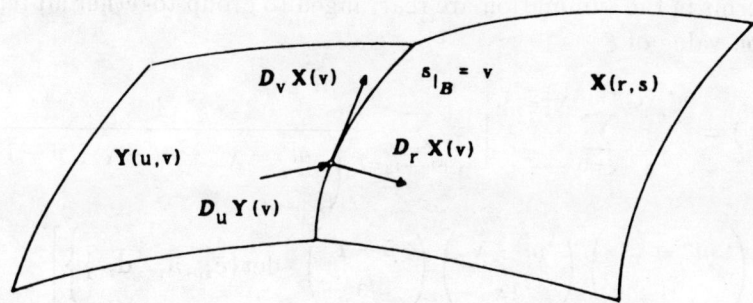

FIG. 10.5 *Derivative vectors of first order of two tensor-product patches joining along a common boundary.*

In order to analyze the coefficient functions $r_{10}(v), s_{10}(v)$ in (10.13) concerning their type, a dual basis system $\{\mathbf{Z}_1(v), \mathbf{Z}_2(v)\}$ corresponding to the basis system $\{\mathbf{D}_r\mathbf{X}(v), \mathbf{D}_v\mathbf{X}(v)\}$ is introduced, i.e.,

$$(10.17) \qquad \begin{aligned} \langle \mathbf{Z}_1(v), \mathbf{D}_r\mathbf{X}(v)\rangle = 1, \quad \langle \mathbf{Z}_1(v), \mathbf{D}_v\mathbf{X}(v)\rangle = 0, \\ \langle \mathbf{Z}_2(v), \mathbf{D}_r\mathbf{X}(v)\rangle = 0, \quad \langle \mathbf{Z}_2(v), \mathbf{D}_v\mathbf{X}(v)\rangle = 1. \end{aligned}$$

Hence the coefficient functions $r_{10}(v), s_{10}(v)$ can be described as scalar products

$$r_{10}(v) = \langle \mathbf{Z}_1(v), \mathbf{D}_u\mathbf{Y}(v)\rangle, \qquad s_{10}(v) = \langle \mathbf{Z}_2(v), \mathbf{D}_u\mathbf{Y}(v)\rangle$$

and by means of vector-algebraic manipulations, the following expressions

$$(10.18) \qquad \begin{aligned} r_{10}(v) &= \frac{\langle \mathbf{D}_u\mathbf{Y}(v) \times \mathbf{D}_v\mathbf{X}(v), \mathbf{D}_r\mathbf{X}(v) \times \mathbf{D}_v\mathbf{X}(v)\rangle}{(\mathbf{D}_r\mathbf{X}(v) \times \mathbf{D}_v\mathbf{X}(v))^2}, \\ s_{10}(v) &= \frac{\langle \mathbf{D}_r\mathbf{X}(v) \times \mathbf{D}_u\mathbf{Y}(v), \mathbf{D}_r\mathbf{X}(v) \times \mathbf{D}_v\mathbf{X}(v)\rangle}{(\mathbf{D}_r\mathbf{X}(v) \times \mathbf{D}_v\mathbf{X}(v))^2}, \end{aligned}$$

are obtained.

Hence, the coefficient functions $r_{10}(v), s_{10}(v)$ prove to be rational functions of the parameter v with a common denominator polynomial (in general, before one starts to cancel possible common linear factors).

10.4.1. Necessary and Sufficient Geometric C^1-Constraints. The coplanarity condition (10.16) provides necessary and sufficient geometric C^1-conditions expressed directly in terms of involved control points [9]. A system of scalar constraints results following the approach to require the vanishing of all Bézier ordinates of the determinant function $\Phi(v)$. Substituting expressions (10.3), (10.6) into condition (10.16) and utilizing multilinearity of determinants, $\Phi(v)$ can be described as

$$\Phi(v) = \sum_{i_1,i_2,i_3} \det\{\overline{\mathbf{c}}_{i_1},\overline{\mathbf{a}}_{i_2},\overline{\mathbf{d}}_{i_3}\} B_{i_1}^{m^*-\chi^*}(v) B_{i_2}^{m-\chi}(v) B_{i_3}^{\overline{p}-1}(v).$$

Introducing a new summation index ξ with $\xi := i_1 + i_2 + i_3$ and using identity (10.9), terms in the summation are rearranged to group together all terms with a common value of ξ

$$\Phi(v) = \sum_{\xi=0}^{m^*-\chi^*+m-\chi+\overline{p}-1} \left[\sum_{i_1+i_2+i_3=\xi} \frac{1}{\left(\dfrac{m^* - \chi^* + m - \chi + \overline{p} - 1}{\xi} \right)} \right.$$

$$\left. \cdot \left\{ \binom{m^* - \chi^*}{i_1} \binom{m - \chi}{i_2} \binom{\overline{p} - 1}{i_3} \cdot \det\{\overline{\mathbf{c}}_{i_1},\overline{\mathbf{a}}_{i_2},\overline{\mathbf{d}}_{i_3}\} \right\} \right]$$

$$\cdot B_{\xi}^{m^*-\chi^*+m-\chi+\overline{p}-1} = 0.$$

The determinant function $\Phi(v)$ vanishes if and only if each of its Bernstein ordinates vanishes. This results in the following Theorem.

THEOREM 10.4.1. *Two adjacent integral Bézier patches* $\mathbf{X}$ *and* $\mathbf{Y}$ *of tensor-product or triangular type joining along a common boundary curve* $\mathbf{B}(v)$ *are tangent plane continuous if and only if the following system of scalar conditions holds for* $\xi = 0(1)m^* - \chi^* + m - \chi + \overline{p} - 1$

$$(10.19) \sum_{i_1+i_2+i_3=\xi} \det\left\{ \binom{m^* - \chi^*}{i_1} \overline{\mathbf{c}}_{i_1}, \binom{m - \chi}{i_2} \overline{\mathbf{a}}_{i_2}, \binom{\overline{p} - 1}{i_3} \overline{\mathbf{d}}_{i_3} \right\} = 0.$$

These scalar conditions are in general independent [9].

Now attention will be focused on developing a system of necessary and sufficient vector-valued geometric C^1-constraints [17]. Please note that $\mathbf{D}_u\mathbf{Y}(v), \mathbf{D}_r\mathbf{X}(v), \mathbf{D}_v\mathbf{X}(v)$ being described by (10.3), (10.6) are vector-valued polynomials of degree $m^* - \chi^*, m - \chi, \overline{p} - 1$, respectively. The rational scalar coefficient functions $r_{10}(v), s_{10}(v)$ in condition (10.13) can be viewed as solutions of an associated linear inhomogeneous equation system.

According to the assumption of regularity along the common boundary, there is at least one non-vanishing component of the cross-product, $\mathbf{D}_r\mathbf{X}(v) \times \mathbf{D}_v\mathbf{X}(v)$ in and nearby an arbitrary but fixed point $\mathbf{B}(\overline{v}), \overline{v} \in [0,1]$ on the common edge. Hence in and near $\mathbf{B}(\overline{v})$ a unique solution to the corresponding scalar equation system of order 2 can be deduced by Cramer's rule, providing information about the lowest upper bounds for numerator and denominator polynomials of $r_{10}(v), s_{10}(v)$. The degree of the common denominator polynomial $t_1(v)$ is not larger than $m - \chi - \bar{p} - 1$, and the degrees of the numerator polynomials $r_1(v), s_1(v)$ are bound by $m^* - \chi^* + \bar{p} - 1$ and $m^* - \chi^* + m - \chi$, respectively. Based on the identity theorem of rational functions, derived properties attached to the degrees of $t_1(v), r_1(v)$, and $s_1(v)$ hold for all $v \in \Re$.

The geometric C^1-joint condition

$$(10.20) \qquad t_1(v)\mathbf{D}_u\mathbf{Y}(v) = r_1(v)\mathbf{D}_r\mathbf{X}(v) + s_1(v)\mathbf{D}_v\mathbf{X}(v)$$

shows a symmetric form: each of the three vector polynomials involved is linked to a scalar polynomial (denoted as form functions).

The following cross-products are now described in a factorized form

$$(10.21) \qquad \begin{aligned} \mathbf{D}_r\mathbf{X}(v) \times \mathbf{D}_v\mathbf{X}(v) &= t_1(v) \cdot \mathbf{P}(v), \\ \mathbf{D}_u\mathbf{Y}(v) \times \mathbf{D}_v\mathbf{X}(v) &= r_1(v) \cdot \mathbf{P}(v), \\ \mathbf{D}_r\mathbf{X}(v) \times \mathbf{D}_u\mathbf{Y}(v) &= s_1(v) \cdot \mathbf{P}(v), \end{aligned}$$

where scalar polynomials $t_1(v), r_1(v), s_1(v)$ are the same as in (10.20) and $\mathbf{P}(v)$ represents an irreducible polynomial vector the components of which have no common scalar factor $\gamma(v)$ with degree $(\gamma) \geq 1$.

The correctness of (10.21) can be deduced from (10.20) and conditions (10.18) by which the following expressions are obtained for the rational coefficient functions $r_{10}(v), s_{10}(v)$:

$$(10.22) \quad r_{10}(v) = \frac{r_1(v)}{t_1(v)} = \frac{\hat{r}_1(v)}{\hat{t}_1(v)} \cdot \frac{\delta_1(v)}{\delta_1(v)}, \qquad s_{10}(v) = \frac{s_1(v)}{t_1(v)} = \frac{\tilde{s}_1(v)}{\tilde{t}_1(v)} \cdot \frac{\varepsilon_1(v)}{\varepsilon_1(v)},$$

with $\gcd(\hat{r}_1(v), \hat{t}_1(v)) = 1, \gcd(\tilde{s}_1(v), \tilde{t}_1(v)) = 1$.

For reasons of regularity the polynomials $r_1(v)$ and $t_1(v)$ do not vanish for $v \in [0, 1]$. A proper orientation of the normal vectors $\mathbf{D}_r\mathbf{X}(v) \times \mathbf{D}_v\mathbf{X}(v)$ and $\mathbf{D}_u\mathbf{Y}(v) \times \mathbf{D}_v\mathbf{X}(v)$ is guaranteed if the form functions $r_1(v)$ and $t_1(v)$ satisfy

$$r_1(v)/t_1(v) > 0 \quad \text{for } v \in [0, 1].$$

The polynomial $s_1(v)$ may have one or several (single or multiple) real zeros for $v \in [0, 1]$, attached to the geometric interpretation that in each corresponding boundary point parametric lines cross to the common edge are tangent continuous.

The polynomial form functions in (10.20) can be expressed in terms of Bernstein polynomials

$$
t_1(v) = \sum_{i=0}^{m-\chi+\overline{p}-1} \tau_{1,i}\, B_i^{m-\chi+\overline{p}-1}(v),
$$

(10.23)
$$
r_1(v) = \sum_{i=0}^{m^*-\chi^*+\overline{p}-1} \rho_{1,i}\, B_i^{m^*-\chi^*+\overline{p}-1}(v),
$$

$$
s_1(v) = \sum_{i=0}^{m-\chi+m^*-\chi^*} \sigma_{1,i}\, B_i^{m-\chi+m^*-\chi^*}(v),
$$

containing shape parameters $\tau_{1,i}, \rho_{1,i}, \sigma_{1,i}$.

Having substituted $t_1(v), r_1(v), s_1(v)$ in (10.20) by (10.23) as well as $\mathbf{D}_u\mathbf{Y}(v), \mathbf{D}_r\mathbf{X}(v), \mathbf{D}_v\mathbf{X}(v)$ by (10.3), (10.6) and using identity (10.9), we obtain, with respect to the independence of the Bernstein basis, a system of linear equations combining difference vectors between control points of the three involved columns and shape parameters.

THEOREM 10.4.2. *A system of geometric C^1-necessary and sufficient continuity conditions for adjacent integral Bézier patches $\mathbf{X}$ and $\mathbf{Y}$ of tensor-product or triangular type joining along a common boundary curve $\mathbf{B}(v)$ is described by*

(10.24)
$$
\sum_{\substack{i=0 \\ i+j=\nu}}^{m-\chi+\overline{p}-1} \sum_{j=0}^{m^*-\chi^*} \binom{m-\chi+\overline{p}-1}{i}\binom{m^*-\chi^*}{j} \tau_{1,i}\,\overline{\mathbf{c}}_j =
$$

$$
\sum_{\substack{i=0 \\ i+j=\nu}}^{m^*-\chi^*+\overline{p}-1} \sum_{j=0}^{m-\chi} \binom{m^*-\chi^*+\overline{p}-1}{i}\binom{m-\chi}{j} \rho_{1,i}\,\overline{\mathbf{a}}_j +
$$

$$
\sum_{\substack{i=0 \\ i+j=\nu}}^{m-\chi+m^*-\chi^*} \sum_{j=0}^{\overline{p}-1} \binom{m-\chi+m^*-\chi^*}{i}\binom{\overline{p}-1}{j} \sigma_{1,i}\,\overline{\mathbf{d}}_j
$$

with $\nu = 0(1)m^* - \chi^* + m - \chi + \overline{p} - 1$.

$\{\tau_{1,i}\}_0^{m-\chi+\overline{p}-1}, \{\rho_{1,i}\}_0^{m^*-\chi^*+\overline{p}-1}, \{\sigma_{1,i}\}^{m-\chi+m^*-\chi^*}$ represent sets of unknown shape parameters to be determined.

10.4.2. Sufficient Geometric C^1-Constraints.

In order to decrease the obviously large number of unknown shape parameters $\tau_{1,i}, \rho_{1,i}, \sigma_{1,i}$ appearing in the system of necessary and sufficient geometric C^1-conditions (10.24) the approach will be to develop sets of sufficient constraints, each corresponding to different degrees of involved form functions and providing different numbers of unknown shape parameters.

Choosing form functions $T_1(v), R_1(v), S_1(v)$ in (10.20) of degree τ_1, ρ_1, σ_1, respectively (these functions are obtained by assuming corresponding common factors of the polynomials $t_1(v), r_1(v), s_1(v)$, which can be cancelled), and a fixed integer λ with range of evaluation

$$\max(m^* - \chi^*, m - \chi, \overline{p} - 1) \le \lambda \le m^* - \chi^* + m - \chi + \overline{p} - 1,$$

we get by analogous deduction the following theorem.

THEOREM 10.4.3. *Sets of geometric C^1-sufficient continuity conditions for adjacent integral Bézier patches* $\mathbf{X}$ *and* $\mathbf{Y}$ *of tensor-product or triangular type joining along a common boundary curve* $\mathbf{B}(v)$ *are represented by*

$$(10.25) \qquad \sum_{\substack{i=0 \\ i+j=\nu}}^{\tau_1} \sum_{j=0}^{m^*-\chi^*} \binom{\tau_1}{i} \binom{m^* - \chi^*}{j} \tau_{1,i}\overline{\mathbf{c}}_j =$$

$$\sum_{\substack{i=0 \\ i+j=\nu}}^{\rho_1} \sum_{j=0}^{m-\chi} \binom{\rho_1}{i} \binom{m - \chi}{j} \rho_{1,i}\overline{\mathbf{a}}_j +$$

$$\sum_{\substack{i=0 \\ i+j=\nu}}^{\sigma_1} \sum_{j=0}^{\overline{p}-1} \binom{\sigma_1}{i} \binom{\overline{p} - 1}{j} \sigma_{1,i}\overline{\mathbf{d}}_j$$

with $\nu = 0(1)\lambda$.

$\{\tau_{1,i}\}_0^{\tau_1}, \{\rho_{1,i}\}_0^{\rho_1}, \{\sigma_{1,i}\}_0^{\sigma_1}$ with $\tau_1 = \lambda - (m^* - \chi^*), \rho_1 = \lambda - (m - \chi), \sigma_1 = \lambda - (\overline{p} - 1)$ describe unknown shape parameters to be suitably determined.

Please note that if $\lambda = m^* - \chi^* + m - \chi + \overline{p} - 1$ the corresponding set of conditions (10.25) coincides with the necessary and sufficient system of geometric C^1-conditions (10.24).

For a detailed discussion of special sufficient cases including all combination variants of neighboring tensor-product and triangular Bézier patches and providing practical construction methods the reader is referred to [21].

10.4.3. Geometric Properties of Special GC^1-Constructions.

In the following considerations indicating geometric properties of special sufficient geometric C^1-conditions we will confine ourselves to the case of two adjacent tensor-product Bézier patches. Only minor modifications are necessary in order to treat combination variants of type triangular–triangular, tensor-product–triangular, or triangular–tensor-product in an analogous way. The involved columns of control points $\mathbf{E}_{0j}, \mathbf{E}_{1j}(j = 0(1)m)$ linked to patch $\mathbf{X}$ are assumed to be given.

Two appropriate sets of form functions

(i) $T_1(v) = 1$, $R_1(v) = \rho_{1,0}$, $\qquad\qquad S_1(v) = \sigma_{1,0}(1 - v) + \sigma_{1,1}v$,
(ii) $T_1(v) = 1$, $R_1(v) = \rho_{1,0}(1 - v) + \rho_{1,1}v$, $S_1(v) = \sigma_{1,0}(1 - v) + \sigma_{1,1}v$,
(10.26)

are introduced, each corresponding to suitably fixed degrees $m^*, m, \overline{p}$:

(i) $m^* = m = \overline{p} = p$ $\qquad\qquad$ (ii) $m^* = \overline{p} = p, m = p - 1$
(10.27)

in directions of the common boundary. Regarding this input we obtain in both cases (i), (ii) just as many vector-valued conditions as unknown Bézier points $\mathbf{F}_{n^*-1,j}$ $(j = 0(1)p)$ of patch $\mathbf{Y}$.

The dependency of the unknown $\mathbf{F}_{n^*-1,j}$ to given Bézier points and suitable shape parameters results from the geometric C^1-condition

$$(10.28) \quad n^* \sum_{j=0}^{m^*} \mathbf{c}_j^{(1)} B_j^{m^*}(v) = R_1(v) n \sum_{j=0}^{m} \mathbf{a}_j^{(1)} B_j^m(v) + S_1(v)\overline{p} \sum_{j=0}^{\overline{p}-1} \mathbf{d}_j B_j^{\overline{p}-1}(v),$$

where $T_1(v)$ has already been set equal to 1. By inserting (10.26), (10.27) in (10.28) and by algebraic calculations, we get two sets of sufficient GC^1-constraints, each yielding a special construction method.

10.4.3.1. GC^1-*construction scheme with geometric interpretation.* Convex combination and barycentric coordinates represent straightforward tools in Farin's geometric C^1-construction idea [10] that is attached to case (i) and provides control points $\mathbf{F}_{n^*-1,j}$ of patch $\mathbf{Y}$ as linear blends

$$\begin{aligned}
\mathbf{F}_{n^*-1,j} &= \left(1 - \frac{j}{p}\right) \left[(1 + \rho_{1,0} + \sigma_{1,0})\mathbf{E}_{0,j} - \sigma_{1,0}\mathbf{E}_{0,j+1} - \rho_{1,0}\mathbf{E}_{1,j}\right] \\
(10.29) & \\
&+ \frac{j}{p}\left[(1 - \sigma_{1,1} + \rho_{1,0})\mathbf{E}_{0,j} + \sigma_{1,1}\mathbf{E}_{0,j-1} - \rho_{1,0}\mathbf{E}_{1,j}\right] \quad (j = 0(1)p).
\end{aligned}$$

Corresponding to corner situations $j = 0, j = p$, equations

$$\begin{aligned}
\mathbf{F}_{n^*-1,0} &= (1 + \rho_{1,0} + \sigma_{1,0})\mathbf{E}_{00} - \sigma_{1,0}\mathbf{E}_{01} - \rho_{1,0}\mathbf{E}_{10}, \\
\mathbf{F}_{n^*-1,p} &= (1 - \sigma_{1,1} + \rho_{1,0})\mathbf{E}_{0p} + \sigma_{1,1}\mathbf{E}_{0,p-1} - \rho_{1,0}\mathbf{E}_{1p},
\end{aligned}$$

allow a geometric interpretation.

With respect to triangle $\triangle_{1,0}$ with vertices $\mathbf{E}_{00}, \mathbf{E}_{01}, \mathbf{E}_{10}$ the corresponding control point $\mathbf{F}_{n^*-1,0}$ has barycentric coordinates $(1 + \rho_{1,0} + \sigma_{1,0}, -\sigma_{1,0}, -\rho_{1,0})$; analogous properties are attached to the Bézier point $\mathbf{F}_{n^*-1,p}$ with respect to triangle $\triangle_{2,p}$. Each of the $p - 1$ translated triangles $\triangle_{1,j}$ with vertices $\mathbf{E}_{0j}, \mathbf{E}_{0,j+1}, \mathbf{E}_{1j}$ originating from triangle $\mathbf{E}_{00}, \mathbf{E}_{01}, \mathbf{E}_{10}$ and forming a first set is appointed to an auxiliary point $\mathbf{H}_{1,j}$ with barycentric coordinates $(1 + \rho_{1,0} + \sigma_{1,0}, -\sigma_{1,0}, -\rho_{1,0})$. In the same way we join auxiliary points $\mathbf{H}_{2,j}$ with barycentric coordinates $(1 - \sigma_{1,1} + \rho_{1,0}, \sigma_{1,1}, -\rho_{1,0})$ to each of the $p - 1$ translated triangles $\triangle_{2,j}$ with vertices $\mathbf{E}_{0,j}, \mathbf{E}_{0,j-1}, \mathbf{E}_{1j}$ being derived from triangle $\mathbf{E}_{0p}, \mathbf{E}_{0p-1}, \mathbf{E}_{1p}$ and composing a second set.

The unknown control points $\mathbf{F}_{n^*-1,j}$ are determined blending linearly together corresponding auxiliary points $\mathbf{H}_{1,j}, \mathbf{H}_{2,j}$ by convex combination according to condition (10.29). See Fig. 10.6.

10.4.3.2. GC^1-*construction scheme compatible to fixed corner positions.* Corresponding to case (ii) control points $\mathbf{F}_{n^*-1,0}, \mathbf{F}_{n^*-1,p}$ attached to corner situ-

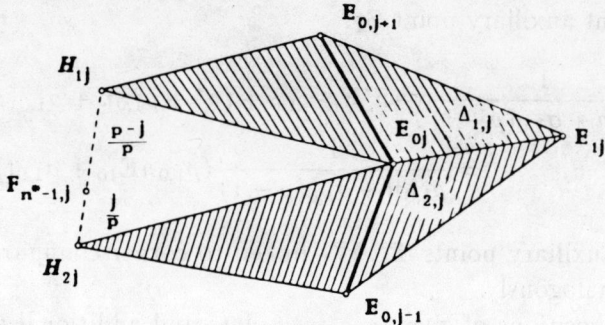

FIG. 10.6 *Geometric interpretation for geometric C^1-construction due to (10.29).*

ations $j = 0, j = p$

$$\mathbf{F}_{n^*-1,0} = \mathbf{E}_{00}\frac{1}{n^*}(n^* + \rho_{1,0}n + \sigma_{1,0}(p-1)) - \mathbf{E}_{10}\frac{1}{n^*}\rho_{1,0}n - \mathbf{E}_{01}\frac{1}{n^*}\sigma_{1,0}(p-1),$$

$$\mathbf{F}_{n^*-1,p} = \mathbf{E}_{0,p-1}\frac{1}{n^*}(n^* + \rho_{1,1}n - \sigma_{1,1}(p-1)) - \mathbf{E}_{1,p-1}\frac{1}{n^*}\rho_{1,1}n$$

$$+ \mathbf{E}_{0,p-2}\frac{1}{n^*}\sigma_{1,1}(p-1),$$

(10.30)

are assumed to have been already fixed appropriately in conformance with design purposes. To guarantee the geometric C^1-joint of the patches $\mathbf{X}$ and $\mathbf{Y}$, the unknown intermediate points $\mathbf{F}_{n^*-1,j}(j = 1(1)p-1)$ have to be evaluated with shape parameters $\rho_{1,0}$, $\rho_{1,1}$, $\sigma_{1,0}$, $\sigma_{1,1}$ compatible to the positions of $\mathbf{F}_{n^*-1,0}$, $\mathbf{F}_{n^*-1,p}$. Hence further attention will be addressed to the problem of determining the unknown shape parameters in (10.30) from given $\mathbf{F}_{n^*-1,0}$, $\mathbf{F}_{n^*-1,p}$.

Please note that the coefficients in each equation (10.30) adding up to one can be viewed as barycentric coordinates and, as a consequence of choosing form functions $R_1(v), S_1(v)$ as linear polynomials, the equations (10.30) are decoupled with respect to occurring shape parameters. Thus the positions of $\mathbf{F}_{n^*-1,0}$ and $\mathbf{F}_{n^*-1,p}$ can be fixed independently in respective tangent planes.

We introduce four auxiliary points, each as an intersection-point of a pair of appropriate lines lying in the same corresponding tangent plane. The auxiliary point $\mathbf{R}_1$, e.g., is provided by the intersection of the line through $\mathbf{E}_{01}, \mathbf{E}_{00}$ and the line through $\mathbf{F}_{n^*-1,0}, \mathbf{E}_{10}$:

$$\mathbf{R}_1 = \frac{1}{n^* + \rho_{1,0}n}(-\sigma_{1,0}(p-1)\mathbf{E}_{01} + (n^* + \rho_{1,0}n + \sigma_{1,0}(p-1))\mathbf{E}_{00})$$

(10.31)

$$= \frac{1}{n^* + \rho_{1,0}n}(n^*\mathbf{F}_{n^*-1,0} + \rho_{1,0}n\mathbf{E}_{10}).$$

Intersection of the line through $\mathbf{F}_{n^*-1,0}, \mathbf{E}_{00}$ and the line through $\mathbf{E}_{10}, \mathbf{E}_{01}$

yields the point auxiliary point S_1:

$$S_1 = \frac{1}{\rho_{1,0}n + \sigma_{1,0}(p-1)} \left(-n^* F_{n^*-1,0} + (n^* + \rho_{1,0}n + \sigma_{1,0}(p-1))E_{00}\right)$$

(10.32)

$$= \frac{1}{\rho_{1,0}n + \sigma_{1,0}(p-1)} \left(\rho_{1,0}n E_{10} + \sigma_{1,0}(p-1)E_{01}\right).$$

Two further auxiliary points T_1, W_1 linked to corner configuration $j = p$ are determined analogously.

With the positions of given control points and additional auxiliary points the ratios

$$\text{ratio } (E_{00}, R_1, E_{01}) \qquad = \frac{-\sigma_{1,0}(p-1)}{n^* + \rho_{1,0}n + \sigma_{1,0}(p-1)},$$

$$\text{ratio } (E_{00}, T_1, F_{n^*-1,0}) \qquad = \frac{-n^*}{n^* + \rho_{1,0}n + \sigma_{1,0}(p-1)},$$

(10.33)

$$\text{ratio } (E_{0,p-1}, T_1, E_{0,p-2}) \quad = \frac{\sigma_{1,1}(p-1)}{n^* + \rho_{1,1}n - \sigma_{1,1}(p-1)},$$

$$\text{ratio } (E_{0,p-1}, W_1, F_{n^*-1,p}) = \frac{-n^*}{n^* + \rho_{1,1}n - \sigma_{1,1}(p-1)},$$

can be evaluated providing a system of linear independent equations for the unknown shape parameters. See Fig. 10.7.

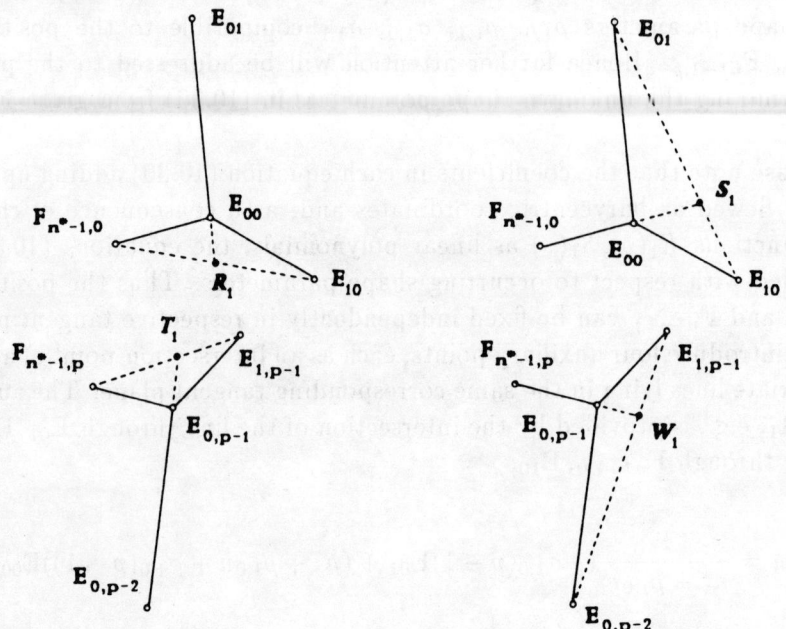

FIG. 10.7 *Evaluation of ratios* (10.33) *attached to corner configurations*
$j = 0, j = p$.

10.5. Geometric C^2-Continuity Conditions

Throughout the following considerations, adjacent Bézier surface patches $\mathbf{X}$ and $\mathbf{Y}$ are assumed to be regular in every point of the common boundary curve $\mathbf{B}(v)$ and to join geometric C^1-continuously along this. The GC^2-condition (10.15) describes the combination of derivative vectors

$$
\text{(10.34)}\quad
\begin{aligned}
&\mathbf{D}_{uu}\mathbf{Y}(v) - [r_{10}(v)]^2\,\mathbf{D}_{rr}\mathbf{X}(v) - 2r_{10}(v)s_{10}(v)\mathbf{D}_{rv}\mathbf{X}(v)\\
&\qquad -[s_{10}(v)]^2\,\mathbf{D}_{vv}\mathbf{X}(v)
\end{aligned}
$$

in linear dependency to the derivatives $\mathbf{D}_r\mathbf{X}(v)$ and $\mathbf{D}_v\mathbf{X}(v)$. Replacing one term, $r_{10}(v)s_{10}(v)\mathbf{D}_{rv}\mathbf{X}(v)$ in (10.34) by

$$
\begin{aligned}
-r_{10}(v)s_{10}(v)\mathbf{D}_{rv}\mathbf{X}(v) &= s_{10}(v)\cdot\left(\tfrac{d[r_{10}(v)]}{dv}\mathbf{D}_r\mathbf{X}(v) + \tfrac{d[s_{10}(v)]}{dv}\mathbf{D}_v\mathbf{X}(v)\right)\\
&\quad - s_{10}(v)\mathbf{D}_{uv}\mathbf{Y}(v) + [s_{10}(v)]^2\mathbf{D}_{vv}\mathbf{X}(v)
\end{aligned}
$$

derived from (10.14), we obtain the geometric C^2-condition (10.15) in an equivalent form,

$$
\begin{aligned}
&\mathbf{D}_{uu}\mathbf{Y}(v) - s_{10}(v)\mathbf{D}_{uv}\mathbf{Y}(v) - [r_{10}(v)]^2\mathbf{D}_{rr}\mathbf{X}(v) - r_{10}(v)s_{10}(v)\mathbf{D}_{rv}\mathbf{X}(v)\\
&= \left(r_{20}(v) - s_{10}(v)\frac{d[r_{10}(v)]}{dv}\right)\cdot\mathbf{D}_r\mathbf{X}(v)\\
&\quad + \left(s_{20}(v) - s_{10}(v)\frac{d[s_{10}(v)]}{dv}\right)\cdot\mathbf{D}_v\mathbf{X}(v),
\end{aligned}
$$

(10.35)

yielding the combination of derivative vectors on the left side in linear dependency to the derivative vectors $\mathbf{D}_r\mathbf{X}(v)$ and $\mathbf{D}_v\mathbf{X}(v)$.

Both representations for the GC^2-condition state corresponding combinations of derivatives of second order as being coplanar to the linear independent derivative vectors $\mathbf{D}_r\mathbf{X}(v), \mathbf{D}_v\mathbf{X}(v)$ which span the tangent plane in each point $\mathbf{B}(\overline{v}), \overline{v} \in [0,1]$ of the boundary. Regarding determinant characterization of Dupin's indicatrix continuity, two adjacent patches join GC^2-continuously in every point of the common boundary if and only if

$$
\begin{aligned}
\Theta_1(v) = \det\{&\mathbf{D}_{uu}\mathbf{Y}(v)(\mathbf{D}_r\mathbf{X}(v)\times\mathbf{D}_v\mathbf{X}(v))^2 - \mathbf{D}_{rr}\mathbf{X}(v)(\mathbf{D}_u\mathbf{Y}(v)\times\mathbf{D}_v\mathbf{X}(v))^2\\
&-2\mathbf{D}_{rv}\mathbf{X}(v)\langle\mathbf{D}_u\mathbf{Y}(v)\times\mathbf{D}_v\mathbf{X}(v), \mathbf{D}_r\mathbf{X}(v)\times\mathbf{D}_u\mathbf{Y}(v)\rangle\\
&-\mathbf{D}_{vv}\mathbf{X}(v)(\mathbf{D}_r\mathbf{X}(v)\times\mathbf{D}_u\mathbf{Y}(v))^2,\ \mathbf{D}_r\mathbf{X}(v),\ \mathbf{D}_v\mathbf{X}(v)\} = 0,
\end{aligned}
$$

(10.36)

or equivalently

$$
\begin{aligned}
\Theta_2(v) = \det\{&\mathbf{D}_{uu}\mathbf{Y}(v)(\mathbf{D}_r\mathbf{X}(v)(\mathbf{D}_r\mathbf{X}(v)\times\mathbf{D}_v\mathbf{X}(v))^2\\
&-\mathbf{D}_{rr}\mathbf{X}(v)(\mathbf{D}_u\mathbf{Y}(v)\times\mathbf{D}_v\mathbf{X}(v))^2\\
&-\mathbf{D}_{uv}\mathbf{Y}(v)\langle\mathbf{D}_r\mathbf{X}(v)\times\mathbf{D}_v\mathbf{X}(v), \mathbf{D}_r\mathbf{X}(v)\times\mathbf{D}_u\mathbf{Y}(v)\rangle\\
&-\mathbf{D}_{rv}\mathbf{X}(v)\langle\mathbf{D}_u\mathbf{Y}(v)\times\mathbf{D}_v\mathbf{X}(v), \mathbf{D}_r\mathbf{X}(v)\times\mathbf{D}_u\mathbf{Y}(v)\rangle,\\
&\mathbf{D}_r\mathbf{X}(v),\ \mathbf{D}_v\mathbf{X}(v)\} = 0
\end{aligned}
$$

(10.37)

holds, where rational coefficient functions $r_{10}(v), s_{10}(v)$ we have been substituted by (10.18) and rational expressions have been avoided by multiplying with $(\mathbf{D}_r\mathbf{X}(v) \times \mathbf{D}_v\mathbf{X}(v))^2$.

The vector-valued GC^2-condition may be rewritten in the form

$$(10.38) \qquad \mathbf{D}_2\mathbf{Y}(v) = \bar{r}_{20}(v)\mathbf{D}_r\mathbf{X}(v) + \bar{s}_{20}(v)\mathbf{D}_v\mathbf{X}(v)$$

using the abbreviations

$$\bar{r}_{20}(v) := r_{20}[t_1(v)]^2, \qquad \bar{s}_{20}(v) := s_{20}(v)[t_1(v)]^2,$$

and

$$(10.39) \qquad \begin{aligned} \mathbf{D}_2\mathbf{Y}(v) &:= (t_1(v))^2\, \mathbf{D}_{uu}\mathbf{Y}(v) - (r_1(v))^2\, \mathbf{D}_{rr}\mathbf{X}(v) \\ &\quad -2r_1(v)s_1(v)\mathbf{D}_{rv}\mathbf{X}(v) - (s_1(v))^2\, \mathbf{D}_{vv}\mathbf{X}(v), \end{aligned}$$

with polynomial functions $t_1(v), r_1(v), s_1(v)$ attached to the GC^1-condition (10.20). Corresponding to the dual basis system $\{\mathbf{Z}_1(v), \mathbf{Z}_2(v)\}$ introduced in §10.4, the coefficient functions $\bar{r}_{20}(v), \bar{s}_{20}(v)$ in (10.38) can be expressed as scalar products

$$\bar{r}_{20}(v) = \langle \mathbf{Z}_1(v), \mathbf{D}_2\mathbf{Y}(v)\rangle, \qquad \bar{s}_{20}(v) = \langle \mathbf{Z}_2(v), \mathbf{D}_2\mathbf{Y}(v)\rangle$$

and vector-algebraic manipulations provide the representations

$$(10.40) \qquad \begin{aligned} \bar{r}_{20}(v) &= \frac{\langle \mathbf{D}_2\mathbf{Y}(v) \times \mathbf{D}_v\mathbf{X}(v), \mathbf{D}_r\mathbf{X}(v) \times \mathbf{D}_v\mathbf{X}(v)\rangle}{(\mathbf{D}_r\mathbf{X}(v) \times \mathbf{D}_v\mathbf{X}(v))^2}, \\ \bar{s}_{20}(v) &= \frac{\langle \mathbf{D}_r\mathbf{X}(v) \times \mathbf{D}_2\mathbf{Y}(v), \mathbf{D}_r\mathbf{X}(v) \times \mathbf{D}_v\mathbf{X}(v)\rangle}{(\mathbf{D}_r\mathbf{X}(v) \times \mathbf{D}_v\mathbf{X}(v))^2}. \end{aligned}$$

Hence the coefficient functions $\bar{r}_{20}(v), \bar{s}_{20}(v)$ are rational functions of the parameter v with a common denominator polynomial (in general, before one starts to cancel possible common linear factors).

10.5.1. Necessary and Sufficient Geometric C^2-Constraints.

Each of the determinant characterizations (10.36), (10.37), with $\Theta_1(v), \Theta_2(v)$ denoting the corresponding determinant functions, provides necessary and sufficient geometric C^2-continuity conditions, expressed directly in terms of involved control points. The approach of requiring the vanishing of all Bézier ordinates of, for instance, $\Theta_1(v)$, yields a system of scalar-valued constraints. In quite an analogous way to §10.4 multi-linearity of determinants is used as well as degree elevation (10.10) and identity (10.9) and a new summation index $\zeta := i_1 + \cdots + i_7$ is introduced to group together all terms with a common value of ζ:

$$\Theta_1(v) = \sum_{\zeta=0}^{t+m^*-2\chi^*+2m-3\chi+2\overline{p}-3} \left[\sum_{i_1+i_2+\dots+i_7=\zeta} \frac{1}{\left(\begin{array}{c} t+m^*-2\chi^*+2m-3\chi+2\overline{p}-3 \\ \zeta \end{array} \right)} \right.$$

$$\cdot \left\{ \left(\begin{array}{c} m^*-2\chi^*+k_1 \\ i_1 \end{array} \right) \left(\begin{array}{c} m-\chi \\ i_2 \end{array} \right) \left(\begin{array}{c} \overline{p}-1 \\ i_3 \end{array} \right) \left(\begin{array}{c} m-\chi \\ i_4 \end{array} \right) \left(\begin{array}{c} \overline{p}-1 \\ i_5 \end{array} \right) \left(\begin{array}{c} m-\chi \\ i_6 \end{array} \right) \left(\begin{array}{c} \overline{p}-1 \\ i_7 \end{array} \right) \right.$$

$$\cdot \det\left([\mathbf{t}_{i_1}]_{k_1} \cdot \langle \overline{\mathbf{a}}_{i_2} \times \overline{\mathbf{d}}_{i_3}, \overline{\mathbf{a}}_{i_4} \times \overline{\mathbf{d}}_{i_5} \rangle, \overline{\mathbf{a}}_{i_6}, \overline{\mathbf{d}}_{i_7} \right)$$

$$- \left(\begin{array}{c} m-2\chi+k_2 \\ i_1 \end{array} \right) \left(\begin{array}{c} m^*-\chi^* \\ i_2 \end{array} \right) \left(\begin{array}{c} \overline{p}-1 \\ i_3 \end{array} \right) \left(\begin{array}{c} m^*-\chi^* \\ i_4 \end{array} \right) \left(\begin{array}{c} \overline{p}-1 \\ i_5 \end{array} \right) \left(\begin{array}{c} m-\chi \\ i_6 \end{array} \right) \left(\begin{array}{c} \overline{p}-1 \\ i_7 \end{array} \right)$$

$$\cdot \det\left([\mathbf{s}_{i_1}]_{k_2} \cdot \langle \overline{\mathbf{c}}_{i_2} \times \overline{\mathbf{d}}_{i_3}, \overline{\mathbf{c}}_{i_4} \times \overline{\mathbf{d}}_{i_5} \rangle, \overline{\mathbf{a}}_{i_6}, \overline{\mathbf{d}}_{i_7} \right)$$

$$- \left(\begin{array}{c} m-\chi-1+k_3 \\ i_1 \end{array} \right) \left(\begin{array}{c} m^*-\chi^* \\ i_2 \end{array} \right) \left(\begin{array}{c} \overline{p}-1 \\ i_3 \end{array} \right) \left(\begin{array}{c} m-\chi \\ i_4 \end{array} \right) \left(\begin{array}{c} m^*-\chi^* \\ i_5 \end{array} \right) \left(\begin{array}{c} m-\chi \\ i_6 \end{array} \right) \left(\begin{array}{c} \overline{p}-1 \\ i_7 \end{array} \right)$$

$$\cdot \det\left([\overline{\mathbf{q}}_{i_1}]_{k_3} \cdot \langle \overline{\mathbf{c}}_{i_2} \times \overline{\mathbf{d}}_{i_3}, \overline{\mathbf{a}}_{i_4} \times \overline{\mathbf{c}}_{i_5} \rangle, \overline{\mathbf{a}}_{i_6}, \overline{\mathbf{d}}_{i_7} \right)$$

$$- \left(\begin{array}{c} \overline{p}-2+k_3 \\ i_1 \end{array} \right) \left(\begin{array}{c} m-\chi \\ i_2 \end{array} \right) \left(\begin{array}{c} m^*-\chi^* \\ i_3 \end{array} \right) \left(\begin{array}{c} m-\chi \\ i_4 \end{array} \right) \left(\begin{array}{c} m^*-\chi^* \\ i_5 \end{array} \right) \left(\begin{array}{c} m-\chi \\ i_6 \end{array} \right) \left(\begin{array}{c} \overline{p}-1 \\ i_7 \end{array} \right)$$

$$\left. \left. \cdot \det\left([\overline{\mathbf{e}}_{i_1}]_{k_3} \cdot \langle \overline{\mathbf{a}}_{i_2} \times \overline{\mathbf{c}}_{i_3}, \overline{\mathbf{a}}_{i_4} \times \overline{\mathbf{c}}_{i_5} \rangle, \overline{\mathbf{a}}_{i_6}, \overline{\mathbf{d}}_{i_7} \right) \right\} \right]$$

$$\cdot B_\zeta^{t+m^*-2\chi^*+2m-3\chi+2\overline{p}-3}.$$

The determinant function $\Theta_1(v)$ vanishes if and only if each of its Bernstein ordinates vanishes. With notation for abbreviation

$$[\widetilde{\overline{\mathbf{t}}}_{i_1}]_{k_1} := [\overline{\mathbf{t}}_{i_1}]_{k_1} \left(\begin{array}{c} m^*-2\chi^*+k_1 \\ i_1 \end{array} \right), \quad [\widetilde{\overline{\mathbf{s}}}_{i_1}]_{k_2} := [\overline{\mathbf{s}}_{i_1}]_{k_2} \left(\begin{array}{c} m-2\chi+k_2 \\ i_1 \end{array} \right),$$

$$[\widetilde{\overline{\mathbf{q}}}_{i_1}]_{k_3} := [\overline{\mathbf{q}}_{i_1}]_{k_3} \left(\begin{array}{c} m-\chi-1+k_3 \\ i_1 \end{array} \right), \quad [\widetilde{\overline{\mathbf{e}}}_{i_1}]_{k_3} := [\overline{\mathbf{e}}_{i_1}]_{k_3} \left(\begin{array}{c} \overline{p}-2+k_3 \\ i_1 \end{array} \right),$$

$$\widetilde{\overline{\mathbf{a}}}_{i_j} := \overline{\mathbf{a}}_{i_j} \left(\begin{array}{c} m-\chi \\ i_j \end{array} \right), (j=2,4,6), \quad \widetilde{\overline{\mathbf{c}}}_{i_j} := \overline{\mathbf{c}}_{i_j} \left(\begin{array}{c} m^*-\chi^* \\ i_j \end{array} \right), (j=2,3,4,5),$$

$$\widetilde{\overline{\mathbf{d}}}_{i_j} := \overline{\mathbf{d}}_{i_j} \left(\begin{array}{c} \overline{p}-1 \\ i_j \end{array} \right), (j=3,5,7),$$

we get the following theorem.

THEOREM 10.5.1. *Two adjacent integral Bézier patches* $\mathbf{X}$ *and* $\mathbf{Y}$ *of tensor-product or triangular type joining tangent-plane continuously along a common boundary curve* $\mathbf{B}(v)$ *are Dupin's indicatrix continuous if and only if the following system of scalar conditions holds for* $\zeta = 0(1)t + m^* - 2\chi^* + 2m -$

$3\chi + 2p - 3$

$$\sum_{i_1+i_2+\ldots+i_7=\zeta} \det \Big\{ [\tilde{\tilde{\mathbf{t}}}_{i_1}]_{k_1} \cdot \langle \tilde{\tilde{\mathbf{a}}}_{i_2} \times \tilde{\tilde{\mathbf{d}}}_{i_3}, \tilde{\tilde{\mathbf{a}}}_{i_4} \times \tilde{\tilde{\mathbf{d}}}_{i_5} \rangle$$

$$-[\tilde{\tilde{\mathbf{s}}}_{i_1}]_{k_2} \cdot \langle \tilde{\tilde{\mathbf{c}}}_{i_2} \times \tilde{\tilde{\mathbf{d}}}_{i_3}, \tilde{\tilde{\mathbf{c}}}_{i_4} \times \tilde{\tilde{\mathbf{d}}}_{i_5} \rangle$$

(10.41)

$$-[\tilde{\tilde{\mathbf{q}}}_{i_1}]_{k_3} \cdot \langle \tilde{\tilde{\mathbf{c}}}_{i_2} \times \tilde{\tilde{\mathbf{d}}}_{i_3}, \tilde{\tilde{\mathbf{a}}}_{i_4} \times \tilde{\tilde{\mathbf{c}}}_{i_5} \rangle$$

$$-[\tilde{\tilde{\mathbf{e}}}_{i_1}]_{k_3} \cdot \langle \tilde{\tilde{\mathbf{a}}}_{i_2} \times \tilde{\tilde{\mathbf{c}}}_{i_3}, \tilde{\tilde{\mathbf{a}}}_{i_4} \times \tilde{\tilde{\mathbf{c}}}_{i_5} \rangle, \tilde{\tilde{\mathbf{a}}}_{i_6}, \tilde{\tilde{\mathbf{d}}}_{i_7} \Big\} = 0.$$

In order to decide on the number of linear independent scalar constraints indicating Dupin's indicatrix continuity, algebraic methods are applied to condition (10.37). Involving control points of the second interior column of each control mesh next to the common boundary $\mathbf{B}(v)$, second-order cross derivatives $\mathbf{D}_{uu}\mathbf{Y}(v), \mathbf{D}_{rr}\mathbf{X}(v)$ can in general be chosen independently of the derivatives of first order $\mathbf{D}_u\mathbf{Y}(v), \mathbf{D}_r\mathbf{X}(v), \mathbf{D}_v\mathbf{X}(v)$ and of mixed derivatives of second order $\mathbf{D}_{uv}\mathbf{Y}(v), \mathbf{D}_{rv}\mathbf{X}(v)$.

Substituting cross-product representations in (10.37) by corresponding expressions (10.22) the determinant formulation takes the form

$$\hat{t}_1(v)(l_1(v))^2(\mathbf{P}(v))^2 \left[\hat{t}_1(v)(t_1(v)\mathbf{D}_{uu}\mathbf{Y}(v) - s_1(v)\mathbf{D}_{uv}\mathbf{Y}(v)) \cdot \mathbf{P}(v) \right.$$

$$\left. -\hat{r}_1(v)(r_1(v)\mathbf{D}_{rr}\mathbf{X}(v) + s_1(v)\mathbf{D}_{rv}\mathbf{X}(v)) \cdot \mathbf{P}(v) \right] = 0.$$

Now because of

$$\gcd(\hat{t}_1(v), \hat{r}_1(v)) = 1$$

$\hat{t}(v)$ must divide

$$(r_1(v)\mathbf{D}_{rr}\mathbf{X}(v) + s_1(v)\mathbf{D}_{rv}\mathbf{X}(v)) \times \mathbf{P}(v),$$

and $\hat{r}_1(v)$ must divide

$$(t_1(v)\mathbf{D}_{uu}\mathbf{Y}(v) - s_1(v)\mathbf{D}_{uv}\mathbf{Y}(v)) \cdot \mathbf{P}(v),$$

implying

$$(r_1(v)\mathbf{D}_{rr}\mathbf{X}(v) + s_1(v)\mathbf{D}_{rv}\mathbf{X}(v)) \cdot \mathbf{P}(v) = \hat{t}_1(v) \cdot L(v) \text{ and}$$

$$(t_1(v)\mathbf{D}_{uu}\mathbf{Y}(v) - s_1(v)\mathbf{D}_{uv}\mathbf{Y}(v)) \cdot \mathbf{P}(v) = \hat{r}_1(v) \cdot M(v),$$

where $L(v), M(v)$ represent scalar polynomials. Hence the geometric C'^2-condition (10.37) can be described in the factorized form

(10.42) $(\hat{t}_1(v))^2 \hat{r}_1(v)(l_1(v))^2(\mathbf{P}(v))^2(L(v) - M(v)) = 0,$

with polynomials already known from the GC^1-condition (10.20) except for $L(v), M(v)$ and the task is to control whether $L(v), M(v)$ represent the same polynomial or not.

The number of conditions to be taken into account depends on the actual degree of the polynomials $L(v), M(v)$. In the following the case, for instance, two adjacent tensor-product patches with $m^* = m = \overline{p} = p$ may be considered. In order to comprehend to general case let us suppose the smallest upper bound for the degrees of both polynomials, which is obviously $3p - 1$. Regarding (10.21), we can conclude that this assumption evidently corresponds to the situation that $\mathbf{P}(v)$ takes its maximal degree of $2p - 1, t_1(v)$ and $r_1(v)$ are constants and $s_1(v)$ is a linear polynomial.

Applied to this situation, DeRose's idea [9] to prove the independence of the resulting conditions can be transferred analogously. Hence, in general, for the considered example, $3p - 1$ constraints are independent of each other as well as of the corresponding $3p - 1$ geometric C^1- constraints.

Now attention will be focused on developing a system of necessary and sufficient vector-valued geometric C^2-constraints. By an analogous deduction to that in §10.4.1, the smallest upper bounds for numerator and denominator polynomials of $r_{20}(v), s_{20}(v)$ are obtained. The degree of the common denominator polynomial $t_2(v)$ is not larger than $m - \chi + \overline{p} - 1$; the degrees of the numerator polynomials $r_2(v), s_2(v)$ are bound by $t + m^* - 2\chi^* + m - 2\chi + 2\overline{p} - 3$ and $t + m^* - 2\chi^* + 2m - 3\chi + \overline{p} - 2$, respectively. The geometric C^2-joint condition

$$(10.43) \qquad t_2(v)\mathbf{D}_2\mathbf{Y}(v) = r_2(v)\mathbf{D}_r\mathbf{X}(v) + s_2(v)\mathbf{D}_v\mathbf{X}(v)$$

shows a symmetric form: each of the three vector polynomials involved is linked to a scalar polynomial (denoted as form functions). The following cross-products are now described in a factorized form

$$(10.44) \qquad \begin{aligned} \mathbf{D}_r\mathbf{X}(v) \times \mathbf{D}_v\mathbf{X}(v) &= t_2(v) \cdot \mathbf{P}(v) = t_1(v) \cdot \mathbf{P}(v), \\ \mathbf{D}_2\mathbf{Y}(v) \times \mathbf{D}_v\mathbf{X}(v) &= r_2(v) \cdot \mathbf{P}(v), \\ \mathbf{D}_r\mathbf{X}(v) \times \mathbf{D}_2\mathbf{Y}(v) &= s_2(v) \cdot \mathbf{P}(v), \end{aligned}$$

where scalar polynomials $t_2(v), r_2(v), s_2(v)$ are the same as in (10.43) and $\mathbf{P}(v)$ represents an irreducible vector polynomial introduced in (10.21).

The correctness of (10.44) can be deduced from (10.43) and (10.40) by which the following expressions are obtained for the rational coefficient functions $\overline{r}_{20}(v), \overline{s}_{20}(v)$:

$$(10.45) \quad \overline{r}_{20}(v) = \frac{r_2(v)}{t_2(v)} = \frac{\hat{r}_2(v)}{\hat{t}_2(v)} \cdot \frac{\delta_2(v)}{\delta_2(v)}, \quad \overline{s}_{20}(v) = \frac{s_2(v)}{t_2(v)} = \frac{\tilde{s}_2(v)}{\tilde{t}_2(v)} \cdot \frac{\varepsilon_2(v)}{\varepsilon_2(v)},$$

where $\gcd(\hat{r}_2(v), \hat{t}_2(v)) = 1$ and $\gcd(\tilde{s}_2(v), \tilde{t}_2(v)) = 1$. Both polynomials $r_2(v), s_2(v)$ may have one or several zeros for $v \in [0, 1]$. Each common zero of the polynomials $s_1(v)$ and $s_2(v)$ for $v \in [0, 1]$ is attached to the geometric interpretation that in the corresponding boundary point parametric lines cross to the common edge, are tangent and curvature-continuous.

From $t_1(v) = t_2(v)$ in (10.44) the conditions result

$$r_1(v)\overline{r}_{20}(v) = r_2(v)r_{10}(v), \qquad s_1(v)\overline{s}_{20}(v) = s_2(v)s_{10}(v)$$

The form functions in (10.43) can be expressed in terms of Bernstein polynomials

$$(10.46) \quad t_2(v)(t_1(v))^2 = \sum_{i=0}^{3(m-\chi+\overline{p}-1)} \tau\tau_i B_i^{3(m-\chi+\overline{p}-1)}(v),$$

$$r_2(v) = \sum_{i=0}^{t+m^*-2\chi^*+m-2\chi+2\overline{p}-3} \rho_{2,i} B_i^{t+m^*-2\chi^*+m-2\chi+2\overline{p}-3}(v),$$

$$s_2(v) = \sum_{i=0}^{t+m^*-2\chi^*+2m-3\chi+\overline{p}-2} \sigma_{2,i} B_i^{t+m^*-2\chi^*+2m-3\chi+\overline{p}-2}(v),$$

$$t_2(v)(r_1(v))^2 = \sum_{i=0}^{2(m^*-\chi^*)+(m-\chi)+3(\overline{p}-1)} \rho\rho_i B_i^{2(m^*-\chi^*)+(m-\chi)+3(\overline{p}-1)}(v),$$

$$t_2(v)r_1(v)s_1(v) = \sum_{i=0}^{2(m^*-\chi^*+m-\chi+\overline{p}-1)} \rho\sigma_i B_i^{2(m^*-\chi^*+m-\chi+\overline{p}-1)}(v),$$

$$t_2(v)(s_1(v))^2 = \sum_{i=0}^{2(m^*-\chi^*)+3(m-\chi)+\overline{p}-1} \sigma\sigma_i B_i^{2(m^*-\chi^*)+3(m-\chi)+\overline{p}-1}(v),$$

containing shape parameters $\tau_{1,i}, \rho_{1,i}, \sigma_{1,i}, \tau_{2,i}, \rho_{2,i}, \sigma_{2,i}$. The parameters $\tau\tau_i, \rho\rho_i, \rho\sigma_i, \sigma\sigma_i$ in (10.46) are already determined by the form parameters appearing in the system of necesssary and sufficient geometric C^1-constraints (10.24).

$$\tau\tau_\nu = \sum_{\substack{i_1=0}}^{m-\chi+\overline{p}-1} \sum_{\substack{i_2=0}}^{m-\chi+\overline{p}-1} \sum_{\substack{i_3=0 \\ i_1+i_2+i_3=\nu}}^{m-\chi+\overline{p}-1}$$

$$\cdot \frac{\dbinom{m-\chi+\overline{p}-1}{i_1}\dbinom{m-\chi+\overline{p}-1}{i_2}\dbinom{m-\chi+\overline{p}-1}{i_3}}{\dbinom{3(m-\chi+\overline{p}-1)}{i_1+i_2+i_3}}$$

$$\cdot \tau_{1,i_1}\tau_{1,i_2}\tau_{1,i_3},$$

(10.47)

$$\rho\rho_\nu = \sum_{\substack{i_1=0}}^{m-\chi+\overline{p}-1} \sum_{\substack{i_2=0}}^{m^*-\chi^*+\overline{p}-1} \sum_{\substack{i_3=0 \\ i_1+i_2+i_3=\nu}}^{m^*-\chi^*+\overline{p}-1}$$

$$\cdot \frac{\dbinom{m-\chi+\overline{p}-1}{i_1}\dbinom{m^*-\chi^*+\overline{p}-1}{i_2}\dbinom{m^*-\chi^*+\overline{p}-1}{i_3}}{\dbinom{2(m^*-\chi^*)+m-\chi+3(\overline{p}-1)}{i_1+i_2+i_3}}$$

$$\cdot \tau_{1,i_1}\rho_{1,i_2}\rho_{1,i_3},$$

$$\rho\sigma_\nu = \sum_{\substack{i_1=0 \\ i_1+i_2+i_3=\nu}}^{m-\chi+\overline{p}-1} \sum_{i_2=0}^{m^*-\chi^*+\overline{p}-1} \sum_{i_3=0}^{m-\chi+m^*-\chi^*}$$

$$\cdot \frac{\begin{pmatrix} m-\chi+\overline{p}-1 \\ i_1 \end{pmatrix} \begin{pmatrix} m^*-\chi^*+\overline{p}-1 \\ i_2 \end{pmatrix} \begin{pmatrix} m-\chi+m^*-\chi^* \\ i_3 \end{pmatrix}}{\begin{pmatrix} 2(m^*-\chi^*)+2(m-\chi)+\overline{p}-1 \\ i_1+i_2+i_3 \end{pmatrix}}$$

$$\tau_{1,i_1}\rho_{1,i_2}\sigma_{1,i_3},$$

$$\sigma\sigma_\nu = \sum_{\substack{i_1=0 \\ i_1+i_2+i_3=\nu}}^{m-\chi+\overline{p}-1} \sum_{i_2=0}^{m-\chi+m^*-\chi^*} \sum_{i_3=0}^{m-\chi+m^*-\chi^*}$$

$$\cdot \frac{\begin{pmatrix} m-\chi+\overline{p}-1 \\ i_1 \end{pmatrix} \begin{pmatrix} m-\chi+m^*-\chi^* \\ i_2 \end{pmatrix} \begin{pmatrix} m-\chi+m^*-\chi^* \\ i_3 \end{pmatrix}}{\begin{pmatrix} 2(m^*-\chi^*)+3(m-\chi)+\overline{p}-1 \\ i_1+i_2+i_3 \end{pmatrix}}$$

$$\cdot \tau_{1,i_1}\sigma_{1,i_2}\sigma_{1,i_3}.$$

Having substituted form functions occurring in (10.43) by (10.46) as well as derivative vectors of first and second order by notation (10.3), (10.4), (10.6) and using identity (10.9) and degree elevation (10.10), we obtain, with respect to the independence of the Bernstein basis, a system of equations combining difference vectors between control points of the five involved columns and shape parameters.

THEOREM 10.5.2. *A system of geometric C^2-necessary and sufficient continuity conditions for adjacent integral Bézier patches* **X** *and* **Y** *of tensor-product or triangular type joining tangent plane continuously along a common boundary curve* **B**(v) *is described by*

$$\sum_{\substack{i=0 \\ i+j=\nu}}^{3(m-\chi+\overline{p}-1)} \sum_{j=0}^{m^*-2\chi^*+k_1} \begin{pmatrix} 3(m-\chi+\overline{p}-1) \\ i \end{pmatrix} \begin{pmatrix} m^*-2\chi^*+k_1 \\ j \end{pmatrix} \tau\tau_i\, [\overline{\mathbf{t}}_j]_{k_1}$$

$$= \sum_{\substack{i=0 \\ i+j=\nu}}^{t+m^*-2\chi^*+m-2\chi+2\overline{p}-3} \sum_{j=0}^{m-\chi} \begin{pmatrix} t+m^*-2\chi^*+m-2\chi+2\overline{p}-3 \\ i \end{pmatrix} \begin{pmatrix} m-\chi \\ j \end{pmatrix} \rho_{2,i}\, \overline{\mathbf{a}}_j$$

$$+ \sum_{\substack{i=0 \\ i+j=\nu}}^{t+m^*-2\chi^*+2m-3\chi+\overline{p}-2} \sum_{j=0}^{\overline{p}-1} \begin{pmatrix} t+m^*-2\chi^*+2m-3\chi+\overline{p}-2 \\ i \end{pmatrix} \begin{pmatrix} \overline{p}-1 \\ j \end{pmatrix} \sigma_{2,i}\, \overline{\mathbf{d}}_j$$

$$+ \sum_{\substack{i=0 \\ i+j=\nu}}^{2(m^*-\chi^*)+m-\chi+3(\overline{p}-1)} \sum_{j=0}^{m-2\chi+k_2} \begin{pmatrix} 2(m^*-\chi^*)+m-\chi+3(\overline{p}-1) \\ i \end{pmatrix} \begin{pmatrix} m-2\chi+k_2 \\ j \end{pmatrix} \rho\rho_i\, [\overline{\mathbf{s}}_j]_{k_2}$$

(10.48)

$$+2 \sum_{\substack{i=0 \\ i+j=\nu}}^{2(m^*-\chi^*+m-\chi+\overline{p}-1)} \sum_{j=0}^{m-\chi-1+k_3} \binom{2(m^*-\chi^*+m-\chi+\overline{p}-1)}{i} \binom{m-\chi-1+k_3}{j} \rho\sigma_i\,[\overline{\mathbf{q}}_j]_{k_3}$$

$$+ \sum_{\substack{i=0 \\ i+j=\nu}}^{2(m^*-\chi^*)+3(m-\chi)+\overline{p}-1} \sum_{j=0}^{\overline{p}-2+k_3} \binom{2(m^*-\chi^*)+3(m-\chi)+\overline{p}-1}{i} \binom{\overline{p}-2+k_3}{j} \sigma\sigma_i\,[\overline{\mathbf{e}}_j]_{k_3}$$

with $\nu = 0(1)t + m^* - 2\chi^* + 2m - 3\chi + 2\overline{p} - 3.$

$\{\tau_{2,i}\}_0^{m-\chi+\overline{p}-1}, \{\rho_{2,i}\}_0^{t+m^*-2\chi^*+m-2\chi+2\overline{p}-3}, \{\sigma_{2,i}\}_0^{t+m^*-2\chi^*+2m-3\chi+\overline{p}-2}$
represent sets of unknown shape parameters to be determined.

10.5.2. Sufficient Geometric C^2-Constraints.

Sets of sufficient constraints, each attached to different degrees of involved form functions and providing different numbers of unknown shape parameters, are developed in order to reduce the evidently large number of unknown shape parameters $\tau_{q,i}, \rho_{q,i}, \sigma_{q,i}$ occurring $(q = 1, 2)$ in the system of necessary and sufficient geometric C^2-conditions (10.48). Linked to the choice of form functions $T_1(v), R_1(v), S_1(v)$ and $T_2(v), R_2(v), S_2(v)$ in (10.43) of degrees τ_1, ρ_1, σ_1 and τ_2, ρ_2, σ_2, respectively (these sets of functions are obtained by assuming corresponding common factors of the set of polynomials $t_1(v), r_1(v), s_1(v)$ and in general other common factors of the set of polynomials $t_2(v), r_2(v), s_2(v)$ which can be cancelled) and of two fixed integers λ, μ with ranges of evaluation

$$\max(m^* - \chi^*, m - \chi, \overline{p} - 1) \le \lambda \le m^* - \chi^* + m - \chi + \overline{p} - 1,$$

$$\max(\Gamma, m - \chi, \overline{p} - 1) \le \mu \le t + m^* - 2\chi^* + 2m - 3\chi + 2\overline{p} - 3,$$

we obtain by analogous deduction the following theorem.

THEOREM 10.5.3. *Sets of geometric C^2-sufficient continuity conditions for adjacent integral Bézier patches* $\mathbf{X}$ *and* $\mathbf{Y}$ *of tensor-product or triangular type joining tangent plane continuously along a common boundary curve* $\mathbf{B}(v)$ *are represented by*

$$\sum_{\substack{i=0 \\ i+j=\nu}}^{\tau_2+2\tau_1} \sum_{j=0}^{m^*-2\chi^*+g_1} \binom{\tau_2+2\tau_1}{i} \binom{m^*-2\chi^*+g_1}{j} \tau\tau_i\,[\mathbf{t}_j]_{g_1}$$

$$= \sum_{\substack{i=0 \\ i+j=\nu}}^{\mu-(m-\chi)} \sum_{j=0}^{m-\chi} \binom{\mu-(m-\chi)}{i} \binom{m-\chi}{j} \rho_{2,i}\,\overline{\mathbf{a}}_j$$

(10.49)

$$+ \sum_{\substack{i=0 \\ i+j=\nu}}^{\mu-(\overline{p}-1)} \sum_{j=0}^{\overline{p}-1} \binom{\mu-(\overline{p}-1)}{i} \binom{\overline{p}-1}{j} \sigma_{2,i}\,\overline{\mathbf{d}}_j$$

$$+ \sum_{\substack{i=0 \\ i+j=\nu}}^{\tau_2+2\rho_1} \sum_{j=0}^{m-2\chi+g_2} \binom{\tau_2+2\rho_1}{i} \binom{m-2\chi+g_2}{j} \rho\rho_i\,[\overline{\mathbf{s}}_j]_{g_2}$$

$$+2 \sum_{\substack{i=0 \\ i+j=\nu}}^{\tau_2+\rho_1+\sigma_1} \sum_{j=0}^{m-\chi-1+g_3} \binom{\tau_2+\rho_1+\sigma_1}{i} \binom{m-\chi-1+g_3}{j} \rho\sigma_i [\overline{\mathbf{q}}_j]_{g_3}$$

$$+ \sum_{\substack{i=0 \\ i+j=\nu}}^{\tau_2+2\sigma_1} \sum_{j=0}^{\overline{p}-2+g_4} \binom{\tau_2+2\sigma_1}{i} \binom{\overline{p}-2+g_4}{j} \sigma\sigma_i [\overline{\mathbf{e}}_j]_{g_4}$$

with $\nu = 0(1)\mu$.

$\{\tau_{1,i}\}_0^{\tau_1}, \{\rho_{1,i}\}_0^{\rho_1}, \{\sigma_{1,i}\}_0^{\sigma_1}$ with $\tau_1 = \lambda-(m^*-\chi^*), \rho_1 = \lambda-(m-\chi), \sigma_1 = \lambda-(\overline{p}-1)$ are shape parameters attached to system (10.25). $\{\tau_{2,i}\}_0^{\tau_2}, \{\rho_{2,i}\}_0^{\rho_2}, \{\sigma_{2,i}\}_0^{\sigma_2}$ with $\tau_2 = \mu - (m^* - 2\chi^* + g_1), \rho_2 = \mu - (m - \chi), \sigma_2 = \mu - (\overline{p} - 1)$ describe shape parameters to be suitably determined.

Please note that if $\mu = t + m^* - 2\chi^* + 2m - 3\chi + 2\overline{p} - 3, \lambda = m^* - \chi^* + m - \chi + \overline{p} - 1$ the corresponding set of conditions (10.49) coincides with the necessary and sufficient system of geometric C^2-conditions (10.48). A detailed discussion of special sufficient cases covering all combination variants of neighboring tensor-product and triangular Bézier patches and providing practical construction methods can be found in [21].

10.5.3. Geometric Properties of Special GC^2-Constructions.

Regarding geometric properties of special sufficient geometric C^2-constraints we will confine ourselves to the case of adjacent tensor-product Bézier patches. With minor modifications combination variants of type triangular–triangular, tensor-product–triangular, or triangular–tensor-product can be handled analogously. The involved columns of control points $\mathbf{E}_{0j}, \mathbf{E}_{1j}, \mathbf{E}_{2j}(j = 0(1)m)$ linked to patch $\mathbf{X}$ are assumed to be given.

Two appropriate sets of form functions

(i) $T_1(v) = 1,$ $R_1(v) = \rho_{1,0},$ $\qquad S_1(v) = \sigma_{1,0}(1 - v) + \sigma_{1,1}v,$
     $T_2(v) = 1,$ $R_2(v) = \rho_{2,0},$ $\qquad S_2(v) = \sigma_{2,0}(1 - v) + \sigma_{2,1}v,$

(ii) $T_1(v) = 1,$ $R_1(v) = \rho_{1,0}(1 - v) + \rho_{1,1}v,$ $S_1(v) = \sigma_{1,0}(1 - v) + \sigma_{1,1}v,$
     $T_2(v) = 1,$ $R_2(v) = \rho_{2,0}(1 - v) + \rho_{2,1}v,$ $S_2(v) = \sigma_{2,0}(1 - v) + \sigma_{2,1}v,$

(10.50)

are introduced, each corresponding to suitably chosen degrees $m^*, m, \overline{p}$:

(i) $m^* = m = \overline{p} = p,$ $\qquad$ (ii) $m^* = \overline{p} = p, m = p - 2,$

(10.51)

in direction of the common boundary. With this input we obtain, in each of both cases (i), (ii), just as many vector-valued conditions as unknown Bézier points $\mathbf{F}_{n^*-1,j}, \mathbf{F}_{n^*-2,j}(j = 0(1)p)$ of patch $\mathbf{Y}$.

The dependency of the unknown $\mathbf{F}_{n^*-1,j}, \mathbf{F}_{n^*-2,j}$ to given Bézier points and suitable shape parameters is provided by the geometric C^1-continuity condition

$$(10.52) \quad n^* \sum_{j=0}^{m^*} \mathbf{c}_j^{(1)} B_j^{m^*}(v) = R_1(v)n \sum_{j=0}^{m} \mathbf{a}_j^{(1)} B_j^m(v) + S_1(v)\overline{p} \sum_{j=0}^{\overline{p}-1} \mathbf{d}_j B_j^{\overline{p}-1}(v),$$

and geometric C^2-continuity condition

$$
n^*(n^* - 1) \sum_{j=0}^{m^*} \left(\mathbf{c}_j^{(1)} - \mathbf{c}_j^{(2)} \right) B_j^{m^*}(v) - [R_1(v)]^2 n(n-1) \sum_{j=0}^{m} \left(\mathbf{a}_j^{(2)} - \mathbf{a}_j^{(1)} \right) B_j^m(v)
$$

$$
-2R_1(v)S_1(v)nm \sum_{j=0}^{m-1} \left(\mathbf{a}_{j+1}^{(1)} - \mathbf{a}_j^{(1)} \right) B_j^{m-1}(v)
$$

$$
-[S_1(v)]^2 \overline{p}(\overline{p} - 1) \sum_{j=0}^{\overline{p}-2} (\mathbf{d}_{j+1} - \mathbf{d}_j) B_j^{\overline{p}-2}(v)
$$

$$
= R_2(v)n \sum_{j=0}^{m} \mathbf{a}_j^{(1)} B_j^m(v) + S_2(v)\overline{p} \sum_{j=0}^{\overline{p}-1} \mathbf{d}_j B_j^{\overline{p}-1}(v),
$$

(10.53)

where $T_1(v) = T_2(v)$ have already been set equal to 1.

By inserting (10.50), (10.51) in (10.52), (10.53) and by algebraic calculations, two sets of sufficient GC^1-/GC^2-constraints result, each describing a special construction method.

10.5.3.1. GC^2-*construction scheme with geometric interpretation.* Convex combination and barycentric coordinates will be used as straightforward construction elements linked to case (i). Above all shape parameters $\sigma_{1,0}, \sigma_{1,1}$ are set to 0. Hence parametric lines of adjacent tensor-product patches cross to the common boundary join tangent continuously. Corresponding conditions in order to evaluate control points $\mathbf{F}_{n^*-1,j}, \mathbf{F}_{n^*-2,j}$ of $\mathbf{Y}$ can be described as follows

(10.54) $\mathbf{F}_{n^*-1,j} = \mathbf{E}_{0j} + \rho_{1,0} \dfrac{n}{n^*} (\mathbf{E}_{0j} - \mathbf{E}_{1j})$ $(j = 0(1)p)$,

$$
\mathbf{F}_{n^*-2,j} = \left(1 - \frac{j}{p} \right)
$$

$$
\left[\left(1 - \sigma_{2,0} \frac{p}{n^*(n^*-1)} - \rho_{2,0} \frac{n}{n^*(n^*-1)} + 2\rho_{1,0} \frac{n}{n^*} + \rho_{1,0}^2 \frac{n(n-1)}{n^*(n^*-1)} \right) \mathbf{E}_{0j} \right.
$$

$$
\left. + \sigma_{2,0} \frac{p}{n^*(n^*-1)} \mathbf{E}_{0,j+1} + \left(\rho_{2,0} \frac{n}{n^*(n^*-1)} - 2\rho_{1,0} \frac{n}{n^*} - \rho_{1,0}^2 \frac{n(n-1)}{n^*(n^*-1)} \right) \mathbf{E}_{1j} \right]
$$

$$
\frac{j}{p} \left[\left(1 + \sigma_{2,1} \frac{p}{n^*(n^*-1)} - \rho_{2,0} \frac{n}{n^*(n^*-1)} + 2\rho_{1,0} \frac{n}{n^*} + \rho_{1,0}^2 \frac{n(n-1)}{n^*(n^*-1)} \right) \mathbf{E}_{0j} \right.
$$

$$
\left. - \sigma_{2,1} \frac{p}{n^*(n^*-1)} \mathbf{E}_{0,j-1} + \left(\rho_{2,0} \frac{n}{n^*(n^*-1)} - 2\rho_{1,0} \frac{n}{n^*} - \rho_{1,0}^2 \frac{n(n-1)}{n^*(n^*-1)} \right) \mathbf{E}_{1j} \right]
$$

$$
+ \rho_{1,0}^2 \frac{n(n-1)}{n^*(n^*-1)} (\mathbf{E}_{2j} - \mathbf{E}_{1j})   (j = 0(1)p),
$$

(10.55)

allowing a straightforward geometric interpretation. See Fig. 10.8.

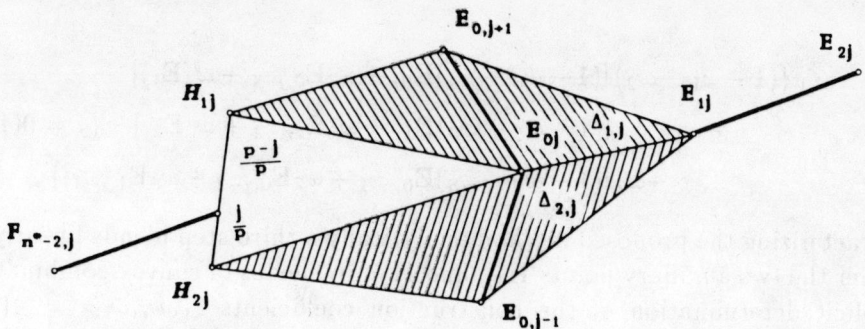

FIG. 10.8 *Geometric interpretation for geometric C^2-construction due to (10.55).*

According to (10.54) $\mathbf{F}_{n^*-1,j}, \mathbf{E}_{0j}$ and $\mathbf{E}_{1j}$ form a collinear set of points. By means of barycentric coordinates an auxiliary point $\mathbf{H}_{1,j}$ is assigned to a corresponding triangle $\triangle_{1,j}$ of a first collection of translated triangles with vertices $\mathbf{E}_{0j}, \mathbf{E}_{0,j+1}, \mathbf{E}_{1j}$ originating from triangle $\mathbf{E}_{00}, \mathbf{E}_{01}, \mathbf{E}_{10}$. Analogously, using barycentric coordinates each of the auxiliary points $\mathbf{H}_{2,j}$ is associated with the matching triangle $\triangle_{2,j}$ of a second collection of translated triangles with vertices $\mathbf{E}_{0j}, \mathbf{E}_{0,j-1}, \mathbf{E}_{1j}$ being derived from triangle $\mathbf{E}_{0p}, \mathbf{E}_{0,p-1}, \mathbf{E}_{1p}$. The construction formula (10.55) blends corresponding auxiliary points by a convex combination and overlaps a succeeding translation in direction $\mathbf{E}_{2j} - \mathbf{E}_{1j}$.

As a generalization with respect to further considerations, shape parameters $\sigma_{1,0}, \sigma_{1,1}$ are assumed to be unequal to 0. Farin's construction scheme yields the column of control points $\mathbf{F}_{n^*-1,j}$. In order to formulate a construction scheme for the Bézier points $\mathbf{F}_{n^*-2,j}$, allowing a geometric interpretation, various choices of triangulations (implying barycentric coordinates) and convex combinations are imaginable attached to the screen of involved control points of $\mathbf{X}$.

Of these possibilities one construction method indicating symmetric properties will be presented. Due to its corresponding geometric view, the proposed scheme consisting of three succeeding steps represents a direct extension of the GC^1-construction idea (10.29). In a first step given Bézier points of $\mathbf{X}$ corresponding to a fixed j are triangulated and auxiliary points $\mathbf{K}_{i,j}(i = 1(1)6)$ are introduced referring to formed triangles $\triangle_{1,i,j}$ by means of barycentric coordinates. Repeating the method of triangulation sets of auxiliary points $\{\mathbf{K}_{1,j}, \mathbf{K}_{2,j}, \mathbf{K}_{3,j}\}$ and $\{\mathbf{K}_{4,j}, \mathbf{K}_{5,j}, \mathbf{K}_{6,j}\}$ describe vertices of the triangles $\triangle_{2,1,j}$ and $\triangle_{2,2,j}$, respectively. Utilizing barycentric coordinates a further auxiliary point $\mathbf{H}_{i,j}(i = 1, 2)$ is joined to each introduced triangle $\triangle_{2,i,j}$. See Fig. 10.9.

According to the construction scheme

$$\mathbf{F}_{n^*-2,j} = (1 - \zeta_1)\{(1 - \psi_1 - \psi_2)[(1 - \psi_3 - \psi_4)\mathbf{E}_{0j} + \psi_3\mathbf{E}_{0,j+1} + \psi_4\mathbf{E}_{1j}]$$

$$(10.56) \qquad + \psi_1 \quad [(1 - \psi_5 - \psi_6)\mathbf{E}_{1j} + \psi_5\mathbf{E}_{1,j+1} + \psi_6\mathbf{E}_{2j}]$$

$$+ \psi_2[(1 - \psi_7 - \psi_8)\mathbf{E}_{0,j+1} + \psi_7\mathbf{E}_{0,j+2} + \psi_8\mathbf{E}_{1,j+1}]\}$$

$$+\zeta_1\left\{(1-\omega_1-\omega_2)\left[(1-\omega_3-\omega_4)\mathbf{E}_{0j}+\omega_3\mathbf{E}_{0,j-1}+\omega_4\mathbf{E}_{1j}\right]\right.$$

$$+\omega_1\quad\left[(1-\omega_5-\omega_6)\mathbf{E}_{1j}+\omega_5\mathbf{E}_{1,j-1}+\omega_6\mathbf{E}_{2j}\right]\quad(j=0(1)p)$$

$$\left.+\omega_2\left[(1-\omega_7-\omega_8)\mathbf{E}_{0,j-1}+\omega_7\mathbf{E}_{0,j-2}+\omega_8\mathbf{E}_{1,j-1}\right]\right\},$$

characterizing the proposed method analytically a third step blends linearly together the two auxiliary points $\mathbf{H}_{1,j}$ and $\mathbf{H}_{2,j}$ by means of convex combination. Explicit determination of the construction coefficients $\zeta_1,\psi_i,\omega_i(i=1(1)8)$ in dependency to shape parameters $\rho_{t,q},\sigma_{t,q}(t,q=1,2)$ and chosen degrees n,n^*,p includes symmetric properties

$$\omega_1=\psi_1,\quad\omega_2=\psi_2,\quad\psi_1=-\psi_2,\quad\omega_6=\psi_6,$$

and yields the relations

$$\zeta_1=\frac{j}{p},\quad\psi_1=\omega_1=-2\beta_0\frac{n}{n^*},\quad\psi_2=\omega_2=2\beta_0\frac{n}{n^*},$$

$$\psi_3=\frac{p}{n^*(n^*-1)}\left(\varphi_0-2\beta_0\frac{n}{p}(n^*-1)-2\gamma_0(n^*-1)-\beta_0\gamma_0 n+2\gamma_0\gamma_1 j\right.$$

$$\left.-\gamma_0^2(p-j-1)\right),$$

$$\omega_3=\frac{p}{n^*(n^*-1)}\left(-\varphi_1-2\beta_0\frac{n}{p}(n^*-1)+2\gamma_1(n^*-1)+\beta_0\gamma_1 n+2\gamma_0\gamma_1(p-j)\right.$$

$$\left.-\gamma_1^2(j-1)\right),$$

$$\psi_4=\frac{n}{n^*(n^*-1)}\left(\varepsilon_0-\beta_0^2(n-1)-\beta_0\gamma_0 p\right),$$

$$\omega_4=\frac{n}{n^*(n^*-1)}\left(\varepsilon_0-\beta_0^2(n-1)+\beta_0\gamma_1 p\right),$$

$$\psi_5=-\frac{1}{2}\gamma_0\frac{p}{n^*-1},\quad\omega_5=\frac{1}{2}\gamma_1\frac{p}{n^*-1},\quad\psi_6=\omega_6=-\frac{1}{2}\beta_0\frac{n-1}{n^*-1},$$

$$\psi_7=\frac{1}{2}\frac{p(p-j-1)}{n(n^*-1)}\frac{\gamma_0^2}{\beta_0},\quad\omega_7=\frac{1}{2}\frac{p(j-1)}{n(n^*-1)}\frac{\gamma_1^2}{\beta_0},$$

$$\psi_8=\frac{1}{2}\gamma_0\frac{p}{n^*-1},\quad\omega_8=-\frac{1}{2}\gamma_1\frac{p}{n^*-1}.$$

(10.57)

10.5.3.2. GC^2-construction scheme compatible to fixed corner positions. Attached to case (ii) control points $\mathbf{F}_{n^*-1,0}$ and $\mathbf{F}_{n^*-2,0},\mathbf{F}_{n^*-1,p}$ and $\mathbf{F}_{n^*-2,p}$ corresponding to corner situations $j=0,j=p$, respectively, are supposed to have already been fixed in a suitable way. To ensure the geometric C^2-joint of the patches $\mathbf{X}$ and $\mathbf{Y}$, the unknown intermediate point

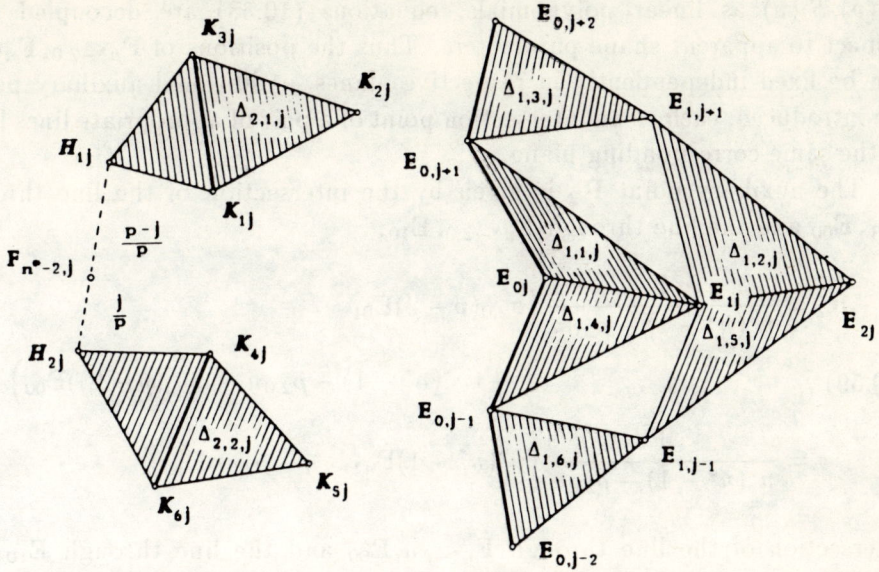

FIG. 10.9 *Geometric interpretation of geometric C^2-construction due to (10.56).*

$\mathbf{F}_{n^*-1,j}, \mathbf{F}_{n^*-2,j}(j = 1(1)p - 1)$ have to be computed with shape parameters $\rho_{i,q}, \sigma_{i,q}, (i = 1, 2; q = 0, 1)$ compatible to the positions of $\mathbf{F}_{n^*-1,0}$ and $\mathbf{F}_{n^*-2,0}, \mathbf{F}_{n^*-1,p}$ and $\mathbf{F}_{n^*-2,p}$. Shape parameters $\rho_{1,0}, \sigma_{1,0}, \rho_{1,1}, \sigma_{1,1}$ involved in the conditions for $\mathbf{F}_{n^*-1,0}, \mathbf{F}_{n^*-1,p}$ are determined applying the method described in §10.4.3. In order to evaluate shape parameters $\rho_{2,0}, \rho_{2,1}, \sigma_{2,0}, \sigma_{2,1}$ compatible to the positions of $\mathbf{F}_{n^*-2,0}, \mathbf{F}_{n^*-2,p}$ appropriate ratios of three different collinear points are considered again. Above all modified control points $\hat{\mathbf{F}}_{n^*-2,0}, \hat{\mathbf{F}}_{n^*-2,p}$ are introduced by translation of the control points $\mathbf{F}_{n^*-2,0}, \mathbf{F}_{n^*-2,p}$:

$$\hat{\mathbf{F}}_{n^*-2,0} := \mathbf{F}_{n^*-2,0} - \hat{\mathbf{F}}_0,$$
$$\hat{\mathbf{F}}_{n^*-2,p} := \mathbf{F}_{n^*-2,p} - \hat{\mathbf{F}}_p,$$

with $\hat{\mathbf{F}}_0, \hat{\mathbf{F}}_p$ incorporating all terms with parameters already known.

As a result of this, the modified control points $\hat{\mathbf{F}}_{n^*-2,0}, \hat{\mathbf{F}}_{n^*-2,p}$ are represented by

$$\hat{\mathbf{F}}_{n^*-2,0} = \frac{1}{n^*(n^*-1)} \Big((n^*(n^*-1) - \rho_{2,0}n - \sigma_{2,0}(p-2))\mathbf{E}_{00} + \rho_{2,0}n\mathbf{E}_{10}$$

$$+ \sigma_{2,0}(p-2)\mathbf{E}_{01} \Big),$$

$$\hat{\mathbf{F}}_{n^*-2,p} = \frac{1}{n^*(n^*-1)} \Big((n^*(n^*-1) - \rho_{2,1}n + \sigma_{2,1}(p-2))\mathbf{E}_{0,p-2} + \rho_{2,1}n\mathbf{E}_{1,p-2}$$

$$- \sigma_{2,1}(p-2)\mathbf{E}_{0,p-3} \Big).$$

(10.58)

The coefficients in each equation (10.58) add up to one and therefore they can be viewed as barycentric coordinates. Because of choosing form functions

$R_2(v), S_2(v)$ as linear polynomials, equations (10.58) are decoupled with respect to apparent shape parameters. Thus the positions of $\hat{\mathbf{F}}_{n^*-2,0}, \hat{\mathbf{F}}_{n^*-2,p}$ can be fixed independently in respective planes. Additional auxiliary points are introduced, each as an intersection point of a pair of appropriate lines lying in the same corresponding plane.

The auxiliary point $\mathbf{R}_2$ is given by the intersection of the line through $\mathbf{E}_{01}, \mathbf{E}_{00}$ and the line through $\hat{\mathbf{F}}_{n^*-2,0}, \mathbf{E}_{10}$:

$$\mathbf{R}_2 = \frac{1}{n^*(n^*-1)-\rho_{2,0}n}\Big(\sigma_{2,0}(p-2)\mathbf{E}_{01}$$

(10.59)
$$+\ (n^*(n^*-1)-\rho_{2,0}n-\sigma_{2,0}(p-2))\mathbf{E}_{00}\Big)$$

$$=\frac{1}{n^*(n^*-1)-\rho_{2,0}n}\Big(n^*(n^*-1)\hat{\mathbf{F}}_{n^*-2,0}-\rho_{2,0}n\mathbf{E}_{10}\Big)$$

Intersection of the line through $\hat{\mathbf{F}}_{n^*-2,0}, \mathbf{E}_{00}$ and the line through $\mathbf{E}_{10}, \mathbf{E}_{01}$ determines the point auxiliary point $\mathbf{S}_2$:

$$\mathbf{S}_2 = \frac{1}{\rho_{2,0}n+\sigma_{2,0}(p-2)}\Big(n^*(n^*-1)\hat{\mathbf{F}}_{n^*-2,0}$$

(10.60)
$$-\ (n^*(n^*-1)-\rho_{2,0}n-\sigma_{2,0}(p-2))\mathbf{E}_{00}\Big)$$

$$=\frac{1}{\rho_{2,0}n+\sigma_{2,0}(p-2)}\Big(\rho_{2,0}n\mathbf{E}_{10}+\sigma_{2,0}(p-2)\mathbf{E}_{01}\Big)$$

Two further auxiliary points $\mathbf{T}_2, \mathbf{W}_2$ are determined analogously attached to corner configuration $j = p$.

With the positions of given control points and additional auxiliary points the ratios

$$\text{ratio}\left(\mathbf{E}_{00}, \mathbf{R}_2, \mathbf{E}_{01}\right) = \frac{\sigma_{2,0}(p-2)}{n^*(n^*-1)-\rho_{2,0}n-\sigma_{2,0}(p-2)},$$

$$\text{ratio}\left(\mathbf{E}_{00}, \mathbf{S}_2, \hat{\mathbf{F}}_{n^*-2,0}\right) = \frac{-n^*(n^*-1)}{n^*(n^*-1)-\rho_{2,0}n-\sigma_{2,0}(p-2)},$$

(10.61)
$$\text{ratio}\left(\mathbf{E}_{0,p-2}, \mathbf{T}_2, \mathbf{E}_{0,p-3}\right) = \frac{-\sigma_{2,1}(p-2)}{n^*(n^*-1)-\rho_{2,1}n+\sigma_{2,1}(p-2)},$$

$$\text{ratio}\left(\mathbf{E}_{0,p-2}, \mathbf{W}_2, \hat{\mathbf{F}}_{n^*-2,p}\right) = \frac{-n^*(n^*-1)}{n^*(n^*-1)-\rho_{2,1}n+\sigma_{2,1}(p-2)},$$

can be calculated providing a system of linear independent equations for the unknown shape parameters. See Fig. 10.10.

10.6. Conclusion

In this paper, the advancements due to Liu and Hoschek concerning the necessary and sufficient conditions for the geometric C^1-joint of adjacent integral

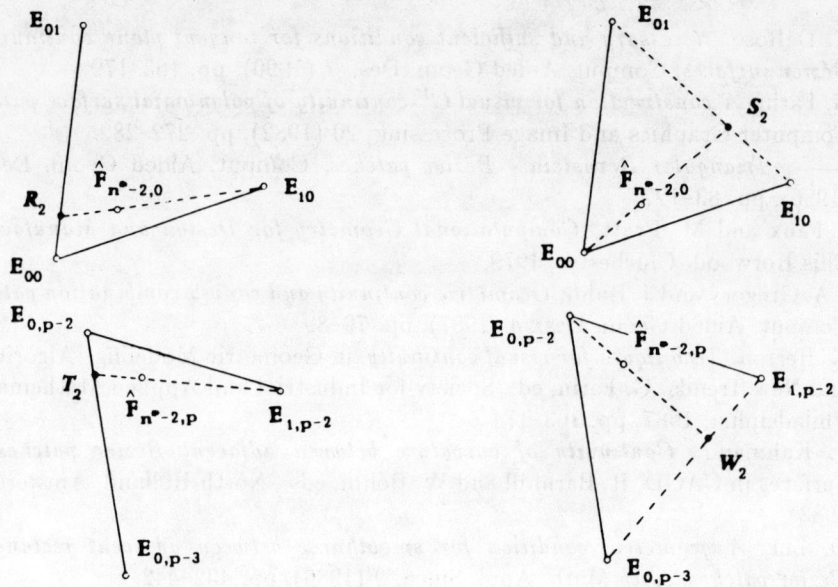

FIG. 10.10 *Evaluation of ratios* (10.61) *attached to corner configurations* $j = 0$, $j = p$.

Bézier surface patches is extended to the case of second-order geometric continuity. With the emphasis being placed on practical applications, special sufficient conditions are derived including form parameters that are to be appropriately chosen for surface modeling purposes. Originating with Farin's geometric C^1-construction, further ideas are proposed for a geometric interpretation linked to a construction scheme that ensures the geometric C'^2-joint of adjacent Bézier surface patches. The conditions developed apply to all the four combination variants between tensor-product and triangular Bézier surface patches.

References

[1] P. Bézier, *Numerical Control–Mathematics and Applications*, John Wiley, New York, 1972.

[2] W. Boehm, G. Farin, and G. Kahmann, *A survey of curve and surface methods in CAGD*, Comput. Aided Geom. Des., 2 (1984), pp. 1–60.

[3] W. Boehm, *Curvature continuous curves and surfaces*, Comput. Aided Geom. Des., 2 (1985), pp. 313–323.

[4] ——, *Smooth curves and surfaces*, in Geometric Modeling: Algorithms and New Trends, G. Farin, ed., Society for Industrial and Applied Mathematics, Philadelphia, 1987, pp. 175–184.

[5] ——, *Visual continuity*, Comput. Aided Geom. Des., 20 (1988), pp. 307–311.

[6] ——, *On G^2-continuity of surfaces*, presented at Topics in CAGD, Jerusalem, June 1988.

[7] S. Cohen, *Beitrag zur steurbaren interpolation von Kurven und Flachen*, Dissertation, TU Dresden, 1982.

[8] W. Degen, *Explicit continuity conditions for adjacent Bézier surface patches*, Comput. Aided Geom. Des., 7 (1990), pp. 181–189.

[9] T. DeRose, *Necessary and sufficient conditions for tangent plane continuity of Bézier surfaces*, Comput. Aided Geom. Des., 7 (1990), pp. 165–179.

[10] G. Farin, *A construction for visual C^1-continuity of polynomial surface patches*, Computer Graphics and Image Processing, 20 (1982), pp. 272–282.

[11] ——, *Triangular Bernstein - Bézier patches*, Comput. Aided Geom. Des., 3 (1986), pp. 83–127.

[12] I. Faux and M. Pratt, *Computational Geometry for Design and Manufacture*, Ellis Horwood, Chichester, 1979.

[13] J.A. Gregory and J. Hahn, *Geometric continuity and convex combination patches*, Comput. Aided Geom. Des., 4 (1987), pp. 79–89.

[14] G. Herron, *Techniques for visual continuity*, in Geometric Modeling: Algorithms and New Trends, G. Farin, ed., Society for Industrial and Applied Mathematics, Philadelphia, 1987, pp. 163–174.

[15] J. Kahmann, *Continuity of curvature between adjacent Bézier patches*, in Surfaces in CAGD, R. Barnhill and W. Böhm, eds., North-Holland, Amsterdam, 1983.

[16] D. Liu, *A geometric condition for smoothness between adjacent rectangular Bézier patches*, Acta Math. Appl. Sinca, 9 (1986), pp. 432–442.

[17] D. Liu and J. Hoschek, *GC^1-continuity conditions between adjacent rectangular and triangular Bézier surface patches*, Comput. Aided Des., 21 (1989), pp. 194–200.

[18] J. Peters, *Local smooth surface interpolation: a classification*, Comput. Aided Geom. Des., 7 (1990), pp. 191–195.

[19] B.R. Piper, *Visually smooth interpolation with triangular Bézier patches*, in Geometric Modeling: Algorithms and New Trends, G. Farin, ed., Society for Industrial and Applied Mathematics, Philadelphia, 1987, pp. 221–233.

[20] M. Vernon, G. Ris, and J. Musse, *Continuity of biparametric surface patches*, Comput. Aided Des., 8 (1976), pp. 267–273.

[21] P. Wassum, *GC^1-and GC^2-Übergangsbedingungen zwischen angrenzenden Rechtecks- und Dreiecks-Bézier-Flächen*, Preprint 1255, Fachbereich Mathematik, Technische Hochschule Darmstadt, 1989.

Index